BALANCING

with clothespins, paper clips, paper beams and simple things

SCIENCE WITH SIMPLE THINGS SERIES

Conceived and written by

RON MARSON

Illustrated by

PEG MARSON

10978 S. Mulino Rd.
Canby, Oregon 97013

ISBN 0-941008-31-2

Library of Congress Catalog Card Number: 81-90443

CONTENTS

INTRODUCTION

TEACHING NOTES

REPRODUCIBLE STUDENT WORKSHEETS

GETTING IT TOGETHER

You hold within your hands a **complete teaching resource.** This book contains 20 reproducible hands-on science lessons together with all necessary information to help you teach each lesson successfully. All you add are the simple materials listed at the bottom of the page.

Look it over. This modest list contains everything you need to teach **every** lesson. Most of the materials you already have. Get the rest from your local supermarket or have your students bring the required items from home.

Each item is **listed in order** of first appearance in the student activities. To start getting it together, begin at the top of this list and work down. Gather everything at once, or collect materials as your students progress through each lesson.

Needed quantities depend on several factors: how you teach, how many students you have and how you organize them into activity groups. The numbers listed by each item correspond to the main teaching strategies in use today. Find the one that suits your teaching style and gather quantities accordingly.

From time to time the teaching notes contain suggestions for additional activities called EXTENSIONS. Materials for these optional experiments are not listed here nor under MATERIALS in the teaching notes. Read instead the extension itself to find out what new materials, if any, are required.

Once you collect the needed materials, place them on an equipment table or on open shelves that are accessible to your students. Items of special value may require a locked cabinet or a special check-out box near the teacher's desk.

Many of the materials you use in this module are used in other TOPS modules as well. As you continue with other TOPS modules and build your inventory, you'll find that gathering materials requires less and less effort!

Q_1

Resource Center
Activity Corner
Parent-Child Activity
Demonstrations

There is enough material so that 1 student or group of students can complete all the activities.

If you multiply Q_1 by 2, then there will be enough materials for two groups to work on the same activity or, perhaps, for three or more groups to simultaneously work on different activities.

Q_2

Individualized Approach

Initial activities require almost as much duplication as the traditional approach. But quantities soon drop off as groups "spread out" within the module, doing different activities at different times.

Students group naturally and informally according to academic or social preferences. Group membership tends to change as slower members fall back into slower groups and faster members move up into faster groups.

Quantities in Q_2 assume a total class size of about 30 students working in 10 groups of 3 each. Modify as necessary to fit your own particular requirements.

Q_3

Traditional Class Lessons

The teacher introduces each lesson to the class as a whole, then everyone does the activity together. Time at the end of the period is reserved for summarizing and reinforcing key concepts.

Quantities in Q_3 again assume a class size of about 30 students working in groups of 3. The numbers are sometimes higher than Q_2 because greater duplication of materials is needed when everyone works simultaneously on the same worksheet.

MATERIALS

$Q_1/Q_2/Q_3$

1 /10/10 pair scissors
1 roll cellophane tape (or substitute masking tape)
1 /10/10 spring-action clothespins
2 /20/20 straight pins (long-sized are best— steel pins if you plan to teach TOPS modules on magnetism.)
1 /10/10 soda bottles or cans— see Teaching Notes 2
1 / 3 / 8 boxes (100's) paper clips— see Teaching Notes 3 (Only 3 boxes needed in Q_3 if you do activity #20 as teacher demo.)
1 / 3 /10 styrofoam cups
1 pkg. modeling clay
3 /24/30 index cards (3x5)
1 pkg. ea. pinto beans, popcorn, lentils, long-grained white rice
3 /12/20 sheets lined notebook paper
1 / 3 /10 pennies
1 / 3 /10 washers — uniform size

SEQUENCING ACTIVITIES

This logic tree shows how all the worksheets in this module tie together. In general, students begin at the trunk of the tree and work up through the related branches. As the diagram suggests, the way to upper level activities leads up from lower level activities.

At the teacher's discretion, certain activities can be omitted or sequences changed to meet specific class needs. The only activities that *must* be completed in sequence are indicated by leaves that are linked vertically with an *open space* in between. In this case the lower activity is a prerequisite to the upper.

When possible, students should complete the worksheets in numerical sequence, from 1 to 20. If time is short, however, or certain students need to catch up, you can use the logic tree to identify concept-related *horizontal* activities. Some of these might be omitted since they serve only to reinforce learned concepts rather than to introduce new ones.

On the other hand, if students complete all the activities at a certain horizontal concept level, then experience difficulty at the next higher level, you might go back down the logic tree to have students repeat specific key activities for greater reinforcement.

For whatever reason, when you wish to make sequence changes, you'll find this logic tree a valuable reference. Parentheses in the upper right corner of each student worksheet allow you this flexibility: they are left blank so you can pencil in sequence numbers of your own choosing.

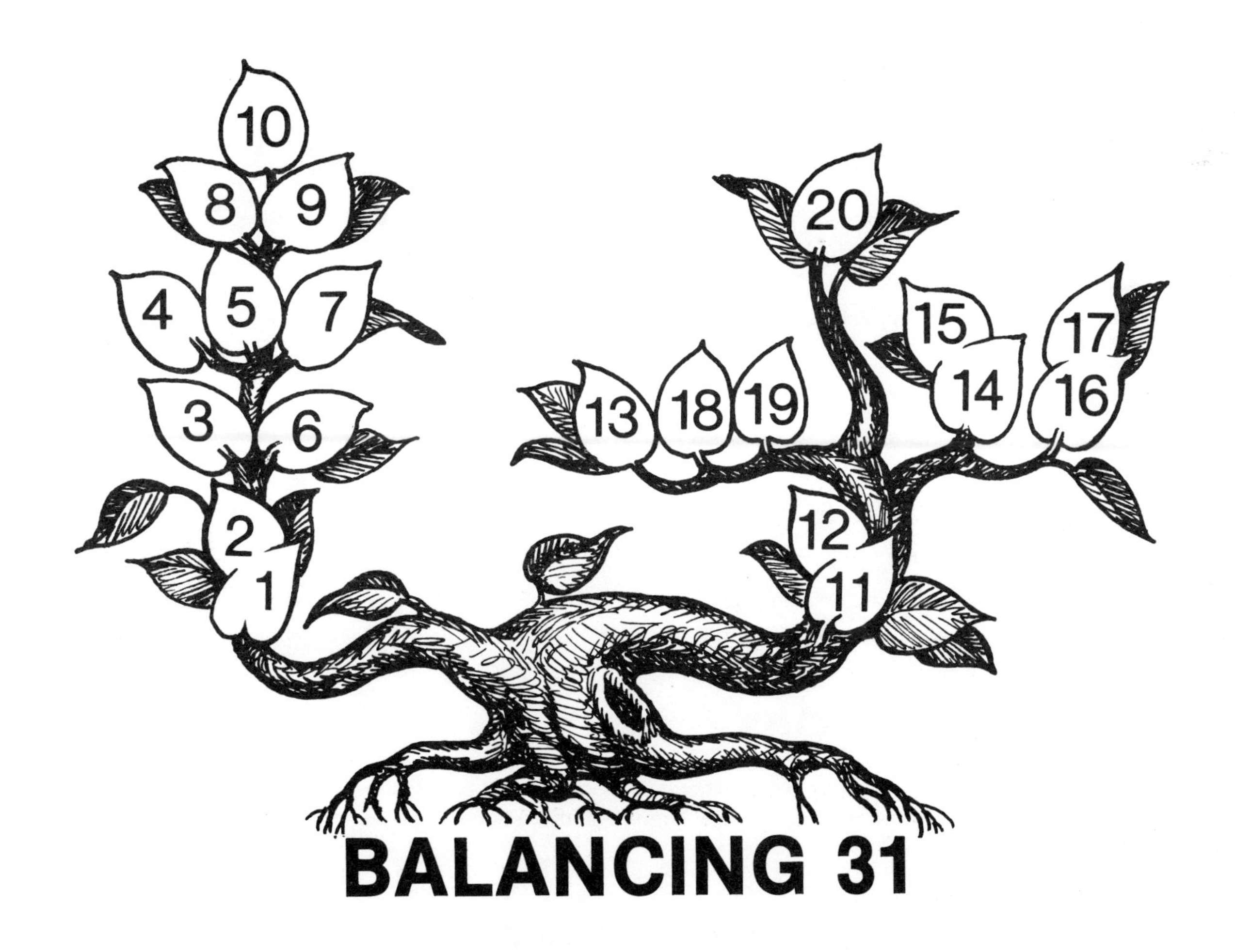

BUILDING AN EFFECTIVE TEACHING STRATEGY

No teaching strategy is totally effective in all classrooms situations. This module is flexibly arranged to adapt to a wide *range* of teaching possibilities. Design your own strategy: select options listed below that best fit your own needs and meet the needs of your students.

A. Classroom Organization

RESOURCE CENTER
Worksheets and science materials are placed in a special resource area. Students come from the classroom to work independently on science activities. Teachers or aides are available to assist students as the need arises.

ACTIVITY CORNER
This operates like the resource center, except a special area is designated *within* the classroom itself. Students come here to do science experiments after their regular class work is completed.

PARENT-CHILD ACTIVITY
A parent, teacher or aide works with one or more students in a tutorial relationship. This may occur during or after school hours or in the home.

TEACHER DEMONSTRATION
The teacher performs experiments in front of the class, inviting occasional student participation. This approach is often used for younger children who do not have sufficient manual dexterity to manipulate the materials.

INDIVIDUALIZED ACTIVITY
Students proceed through the worksheets at their own pace. Those working on the same activity informally group together, reducing substantially the total number of experiments going on in the classroom. The teacher acts as a learning supervisor, responding to questions and problems as they arise within the context of class activity. After the most advanced students complete all the worksheets, the class moves on to a new module *together.*

TRADITIONAL CLASS TOGETHERNESS
Each activity constitutes a specific lesson to be completed during a specified time frame. The teacher introduces the activity to all the students together, then breaks the class into managable lab groups that each do the same experiment. A class discussion sometimes follows to summarize key concepts and provide lesson closure.

B. Reproduction of Activity Sheets

OVERHEAD PROJECTION
Place each worksheet directly on an opaque projector or prepare a transparency.

ACTIVITY CARDS
Make 2 or 3 photocopies of each worksheet. Plasticize them to make durable full page activity cards. File these in an activity folder or display them on a bulletin board or wall.

WORKSHEETS
Duplicate enough copies to provide each student with a worksheet. These can be photocopied directly or thermofaxed onto a master ditto, then run off on a spirit duplicator. Distribute copies to students directly. Or place each set in a separate folder and file them in a box so students can use them as needed.

C. Evaluation

PASS / NO-PASS CHECKPOINTS
Daily write-ups are evaluated by the student and teacher together *in class*. If the student demonstrates reasonable effort commensurate with ability, the write-up is simply checked off, either in a grade book or on a progress chart attached to each student's personal assignment folder kept on file in class.

GRADED ASSIGNMENTS
Write-ups are handed in by each student as completed, graded by a teacher or an aide and then returned to the student.

QUIZES
The teacher gives a quiz (written or oral) after each activity. Questions for the quiz are taken from the "Evaluation" sections in the teaching notes.

INFORMAL OBSERVATION
The teacher takes mental note of active participators who work to capacity and of inactive onlookers who waste time. Grades are awarded accordingly.

EXAMS
An exam given to all students at the same time covers key concepts from activities that all students have completed and reviewed. Questions come from the "Evaluation" sections in the teaching notes.

Among the teaching options listed above, we recommend the combination A5 - B3 - C1 - C4 - C5. This approach combines elements from two opposite teaching strategies in a most effective way: it allows for individual differences while maintaining traditional class togetherness.

Would such an approach work in your own classroom? The "Dairy of a Teacher" which follows may help you visualize the answer. It is based on my own classroom experience.

DIARY OF A TEACHER

THE DAY BEFORE

Tomorrow is the first day of school. With anxiety and anticipation you check to see that everything in your classroom is in good order.

You have already duplicated 30 copies each of the first few lessons in this TOPS module. They lay on your desk, each in a manilla folder marked with large numbers 1, 2 and 3. You make a mental note to bring a large box and a brick to school tomorrow. These will prop the folders upright like a vertical file, and help keep you organized as you add additional folders.

You wonder how you ever managed without manilla folders. Already you have printed each student's name on a fresh new assignment folder and stapled a sheet of graph paper to the inside cover to track each student's progress. When they arrive tomorrow, you will surprise them with a worksheet, a file folder and the simple instruction, "Get busy". You've already laid out the necessary materials on a table in the back of your room. You smile inside yourself; you haven't felt quite this prepared in years.

DAY 10

Your class has been humming now for 10 straight days. Perhaps not humming: buzzing more aptly describes the state of orderly confusion. Students have questions and problems to be sure. (You wish they would at least *read* the instructions before running to you for an explanation.) Still, the worksheets provide a firm sense of direction. Students know where they are and understand where they need to go.

Now that students understand your system, they come to class and get straight to work on what they were doing the day before. Just before lunch they tend to quit early, but at other times you have to pry them away from their experiments. You tell the slower ones to assign themselves homework to catch up and it seems to be working!

The assignment folders work well too. Students point with pride at the growing list of check-points you have marked off on their graph paper progress charts. As their folders expand so does their self confidence.

DAY 15

Today 2 groups of students who seem to be racing each other have completed all 20 activities. The bulk of your class remains 3 to 6 activities behind, with a few stragglers plus the new kid bringing up the rear.

You announce that individualized worksheet activity will end in 2 days. The most advanced students seem eager to work on several experiments of their own. You can follow that up with an "Extension" activity if time allows. You help the slow ones catch up by assigning three key-concept activities while skipping the rest.

There is a frenzy of activity as students rush to meet your deadline. They know that part of their grade is determined by the total number of activities they complete.

DAY 18

Today you kick back and relax. You have assigned several students to give reports on their original investigations. The rest of the period will be taken up with a film. For tomorrow you've planned a blackboard review of major module concepts. Then on Friday you'll finish off with an exam.

The kids are already bugging you about grades and asking what they will be studying next. You decide to give them a 3-part grade weighted equally on pace (number of lessons completed), attitude, and the exam. As to what they'll study next, you can't decide. Perhaps you'll let *them* decide.

Its already the fourth week into the school year and you don't even feel the strain. Activity-centered teaching seems so natural and easy. You respond to questions that kids have instead of the other way around. You ask yourself why you never taught this way before.

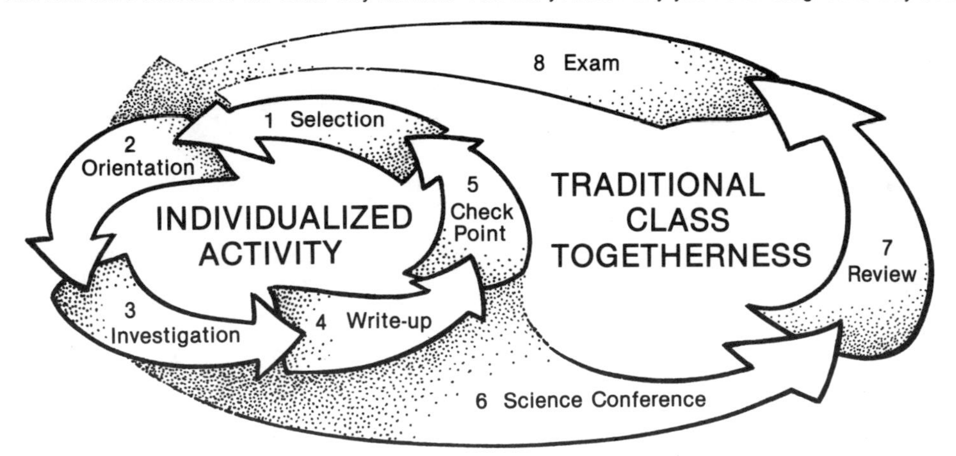

1. SELECTION. Students generally select worksheets in the order you specify. They should be allowed to skip a task that is not challenging, however, or repeat a task with doubtful results. When possible, encourage students to do original investigations that go beyond or replace particular activites.

2. ORIENTATION. Good students will simply read worksheet instructions and understand what to do. Others will require further verbal interpretation. Identify poor reader in your class. When they ask, "What does this mean?" they may be asking in reality, "Will you please read these instructions aloud?"

3. INVESTIGATION. Students observe, hypothesize, predict, test and analyze, often following their own experimental strategies. The teacher provides assistance where needed and the students help each other. When necessary, the teacher may interrupt individual activity to discuss problems or concepts of general class interest.

4. WRITE-UP. Worksheets ask students to explain the how and why of things. Answers should be brief and to the point. Students may accelerate their pace by completing these out of class.

5. CHECK POINT. The student and teacher evaluate each write-up together on a pass/no-pass basis. If the student has made reasonable effort consistent with individual ability, the write-up is checked off on a progress chart and included in the student's personal assignment folder kept on file in class.

6. SCIENCE CONFERENCE. After individualized activity has ended, students come together to discuss experiments of general interest. Those who did original investigations give brief reports. Slower students learn about the later activities completed only by faster students. Newspaper articles are read that relate to the topic of study. The conference is open to speech making, debate, films, celebration, whatever.

7. REVIEW. Important concepts are discussed and applied to problem solving in preparation for the module exam.

8. EXAM. Evaluation questions are written in the teaching notes that accompany each activity. They determine if students understand key concepts developed in the worksheets. Students who finish the test early begin work on the first activity in the next new module.

LONG-RANGE OBJECTIVES

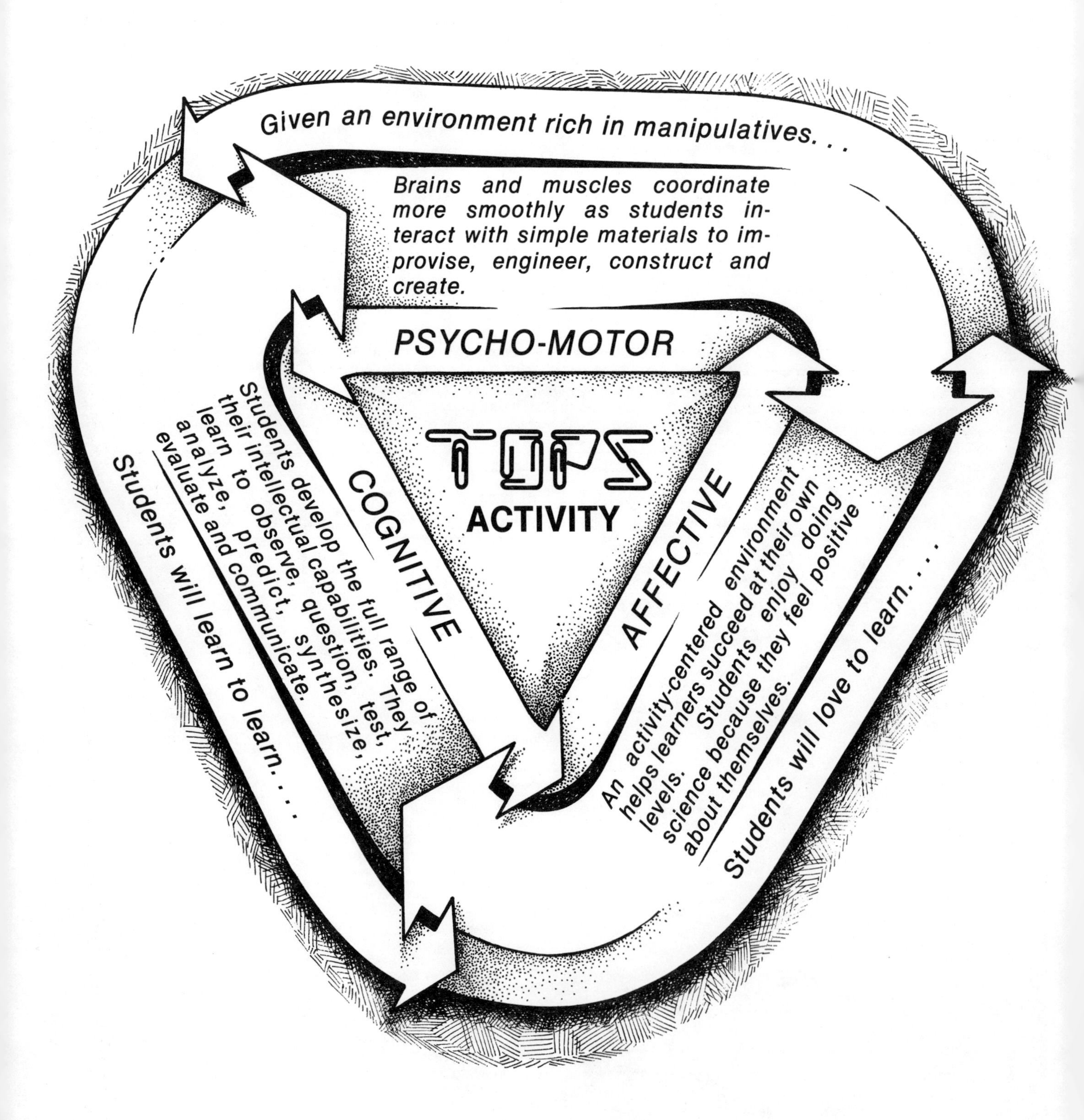

GAINING A WHOLE PERSPECTIVE

Science is an interconnected fabric of ideas woven into broad and harmonious patterns. Use "Extensions" in the teaching notes plus the outline presented below to help your students grasp the big ideas–to appreciate the fabric of science as a unified whole.

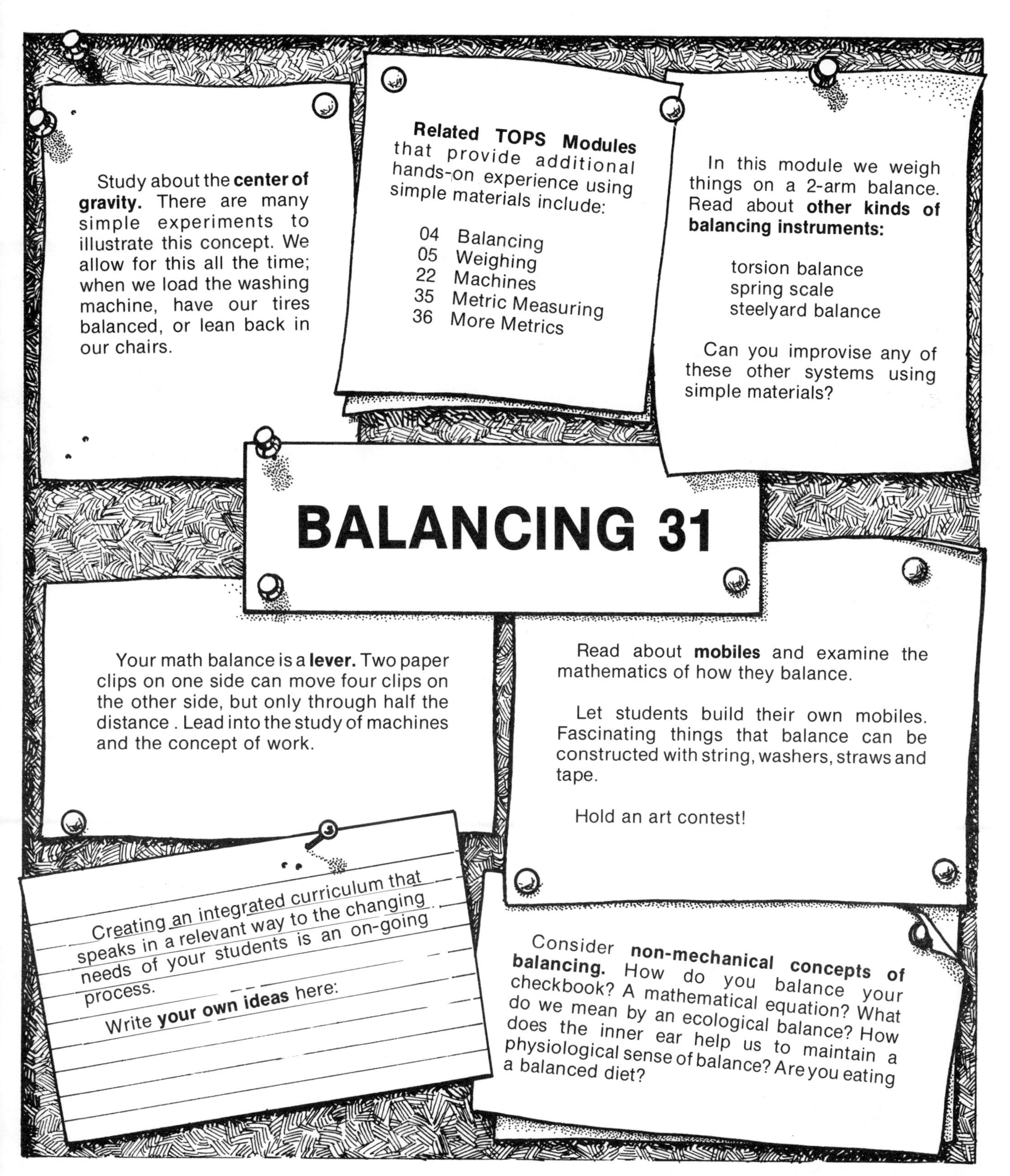

TEACHING NOTES
For Activities 1-20

Task Objective (TO) fold a paper beam to use in a math balance.

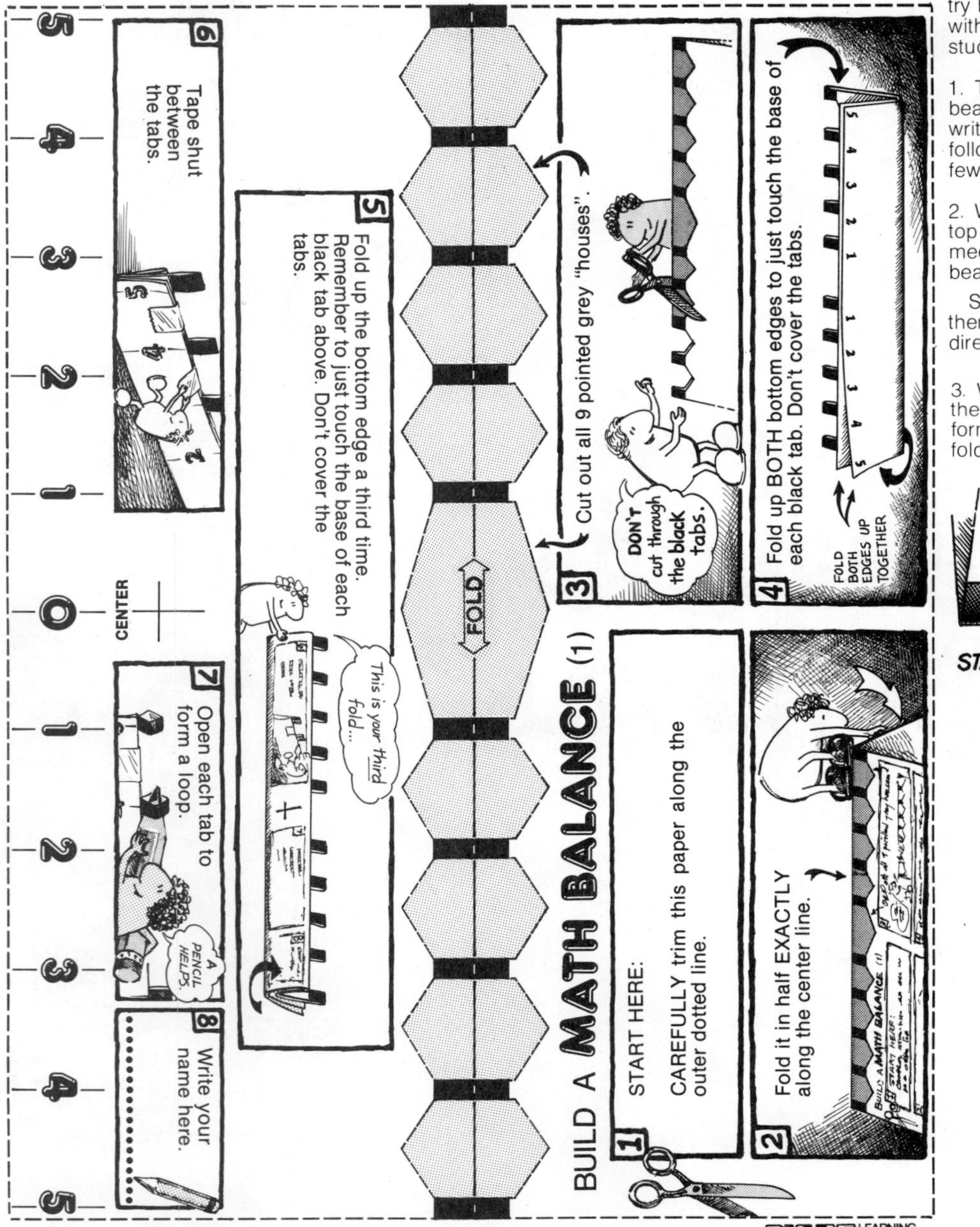

TEACHING NOTES 1

In this activity and the next, your students will build their own math balances. Before they start, try building one yourself. This will familiarize you with the directions, and provide a model for your students to follow.

1. This worksheet folds into an actual balance beam as students complete instructions 1-8 written upon it. Students who begin at step 1 and follow the directions in sequence will experience few difficulties in completing their math beams.

2. When folded together along the guide line, the top and bottom edges of the paper don't quite meet. These edges will "creep" together as the beam is folded over in step 4, and again in step 5.

Some students may fold the paper inward and thereby cover the directions. The illustration directs students to fold their worksheets outward.

3. When you cut out the grey houses, you form the black tabs. The easiest way to cut well-formed, uniform tabs is to always begin at the fold, cutting upward toward each apex.

BEST WAY TO CUT:

END END

START START START

Students who cut around each house in one continuous motion, tend to produce tabs that are less even.

MORE DIFFICULT WAY TO CUT:

START..... END START...... END

Tabs that are cut too narrow, or even sliced off, can always be patched back together with cellophane tape and then recut.

4. *Both* bottom edges fold up to the base of the tabs. Students have a strong tendency to fold up just 1 edge, contrary to the instructions.

5. This is the third and final doubling. Watch out for students who skip this step, ending up with a balance beam that is twice as wide as it should be.

8. A completed balance beam should look something like this.

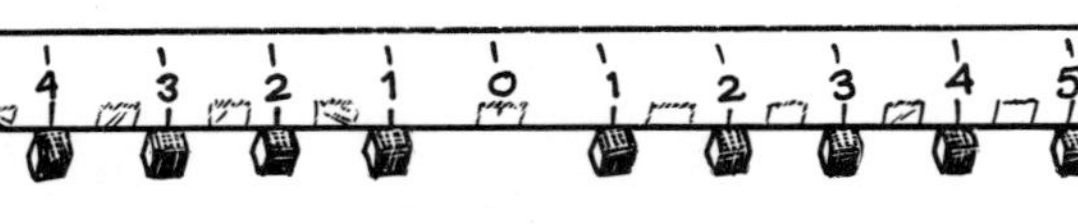

The beams your students make may not appear so well put together. But don't despair. Despite bad cuts and crooked folds, most beams still work reasonably well.

Evaluation

Is the paper well folded so the edges match more or less evenly? Are the tabs properly cut and pushed to open loops? (Tabs that are cut too wide, with white showing along the cut edges, should be trimmed.)

Materials

- ☐ A pair of scissors.
- ☐ Cellophane tape.

(TO) complete construction of the math balance.

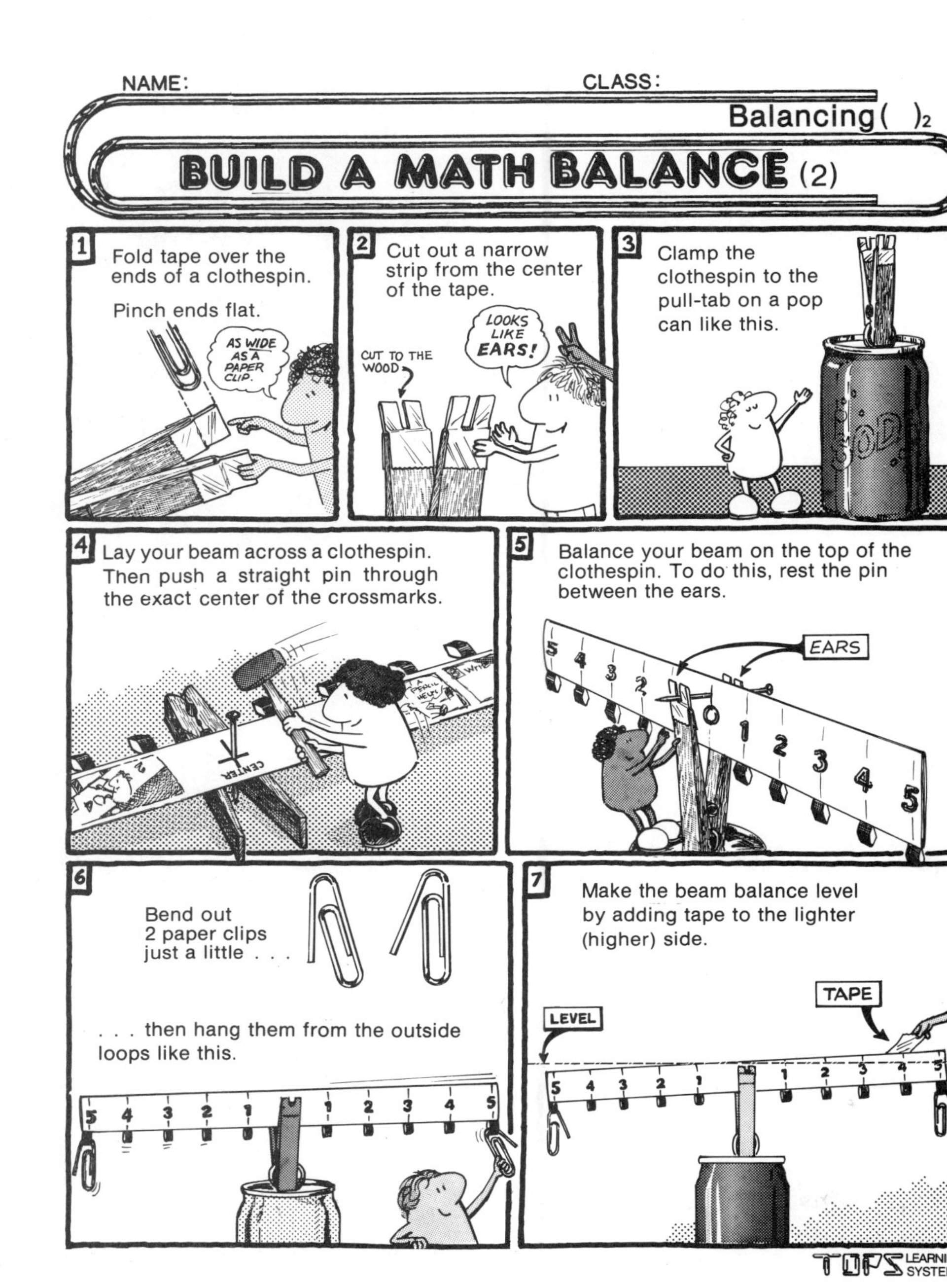

1. The tape must not extend too far beyond the ends of the clothespin. Otherwise it will curl and interfere with the free motion of the beam in between.

2. If your scissors are blunt, you may have to scrape off this narrow piece of tape rather than cut it off. It's best to remove it completely so the pivot pin can rest directly on solid wood in step 5.

3. You can substitute soda bottles for pop cans to serve as balance bases. Some bottles may have openings with the right diameter, so a clothespin will fit snugly inside. Others may have a mouth that is too large. You can make it narrower by sticking tape around the inside of the rim.

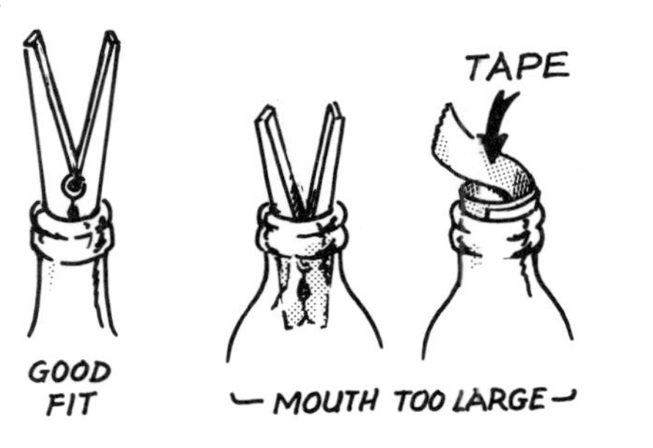

4. It is important to push the pin through the ***exact*** center of the cross mark. Less coordinated students may need help on this step. Propping the beam on the clothespin provides space for the pin to penetrate once it is pushed through the beam. Students who attempt this step while holding the beam in their fingers risk getting a pin prick.

5. Students who can't get their beams to balance may have placed them upside down between the clothespin. Check to see that the black tabs are pointing ***down.***

In general, the beam becomes more stable as you place the pin higher and more unstable (but more sensitive) as you place the pin lower.

RECOMMENDED POSITION — MORE STABLE — MORE SENSITIVE — UNSTABLE

The intersection of the cross marks is only a recomended compromise between stability and sensitivity. Make adjustments up or down the vertical line as necessary.

6-7. You can center the beam without the paper clips, but it may come to rest more slowly, drifting from side to side, past its equilibrium position. Identical paper clips (they must weigh the same!) lower the beam's overall center of gravity, making it balance more readily.

A good way to increase the overall stability of the balance base is to fill the can (or bottle) 1/3 full with gravel, sand or even dirt. You can also hold the clothespin more tightly to the can (or bottle mouth) by winding it with tape.

Evaluation

Is the pin pushed through the beam exactly on center? Does the beam swing freely, returning each time to a level position?

Materials

☐ Tape.
☐ A straight pin.
☐ A spring-action clothespin.
☐ A pry-top soda bottle or soda can. See note 3 above.
☐ Paper clips of uniform size and weight.

(TO) get acquainted with a math beam. To diagram various ways that paper clips balance on the beam.

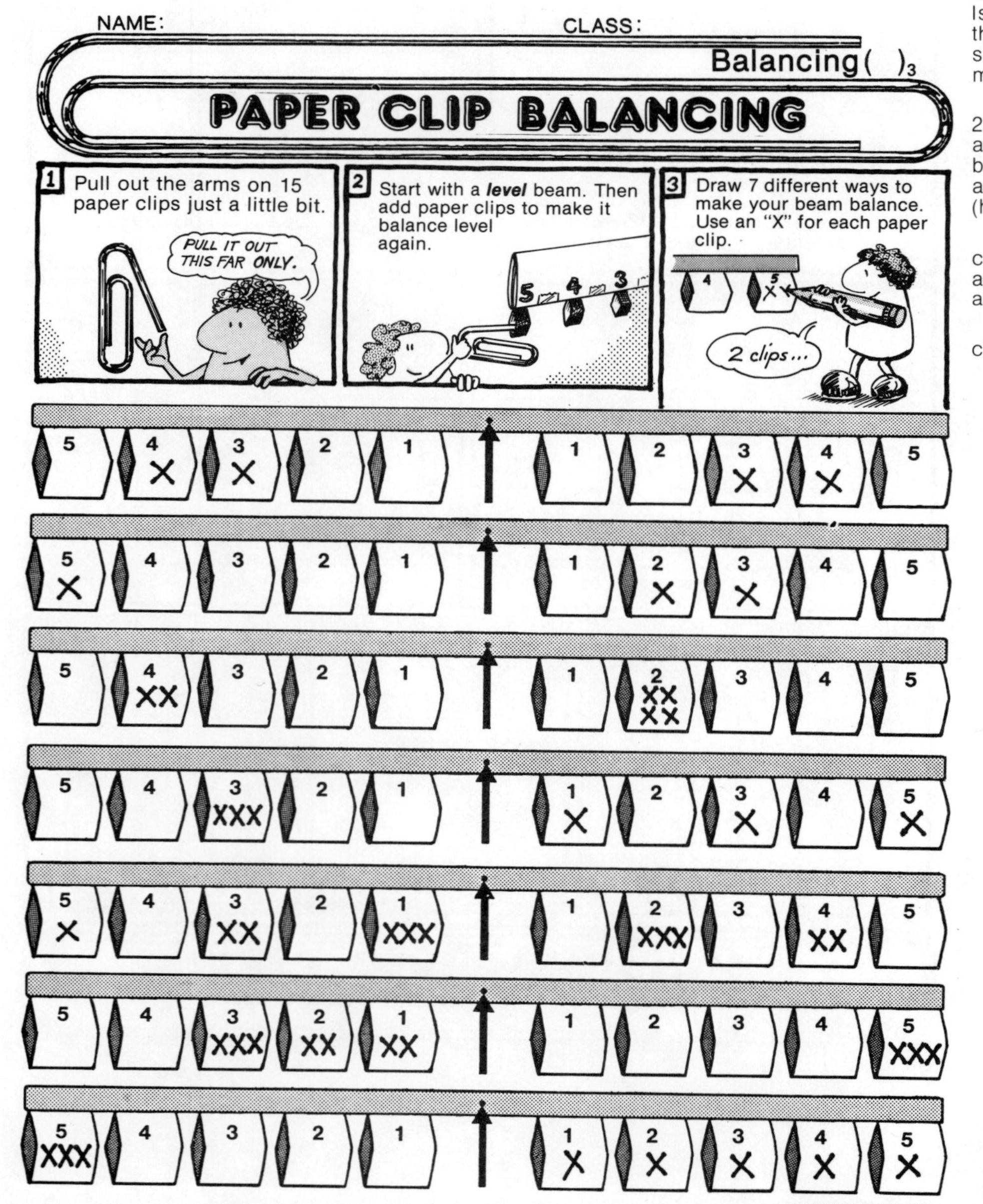

TEACHING NOTES 3

1. Students tend to exaggerate this operation. Remind them to pull out the "arm" of the paper clip *just a little*.

All paper clips used in this module ***must*** have uniform size and weight. Otherwise the balance will not function in a simple mathematical way. Isolate odd brands of paper clips and banish them from your classroom! Only a single kind should circulate while you are teaching this module.

2-3. As a matter of procedure, students should always check first to see that their beams balance level ***before*** adding paper clips. Correct any imbalance by adding tape to the lighter (higher) side.

Then hang paper clips from the black tabs in combinations that make the beam balance level again. Mark the position of each clip by drawing an "X" in the appropriate box.

Paper clips may be hung from the beam in clusters or chains.

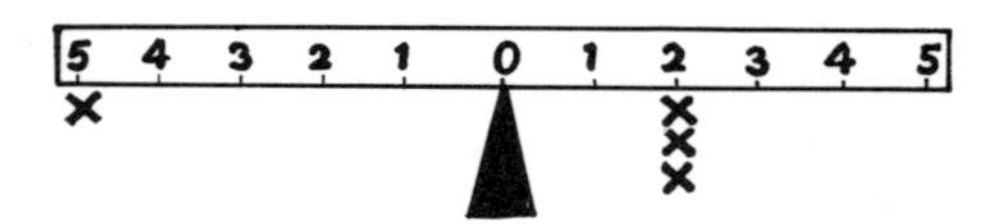

3. There are many different ways to balance the beam with paper clips. Answers listed here represent only a few of the many possibilities.

Some students may limit their investigation to a symmetrical distribution of paper clips on the beam (see first example at top). Encourage them to make an ***uneven*** number of paper clips balance.

Variations on the "X" symbol used for paper clips are fine. Artists in your class may prefer to draw real paper clips! Do not, however, allow students to substitute actual numbers. If three paper clips belong in a specific box, for example, they may write "XXX" but not "3".

Your more advanced students may discover, with great triumph, that their math balances can add and multiply! This becomes apparent in the next two activities.

The math balances will last longer if students are responsible for their own. To this end, they might be assigned a specific place to keep their assembled balances and paper clips.

Evaluation

Q. Use your math balance (if necessary) to decide if this beam balances.

5 4 3 2 1 0 1 2 3 4 5
X (at left 5); XXX (at right 2)

A. No. The beam tilts down to the right.

Materials

☐ The math balance constructed in activities 1 and 2.
☐ Paper clips of uniform size and weight, about 15 per math balance. Use a standard size about this large. (See note 1 above.)

(TO) understand that paper clips add up to equal sums on each arm of a *balanced* beam.

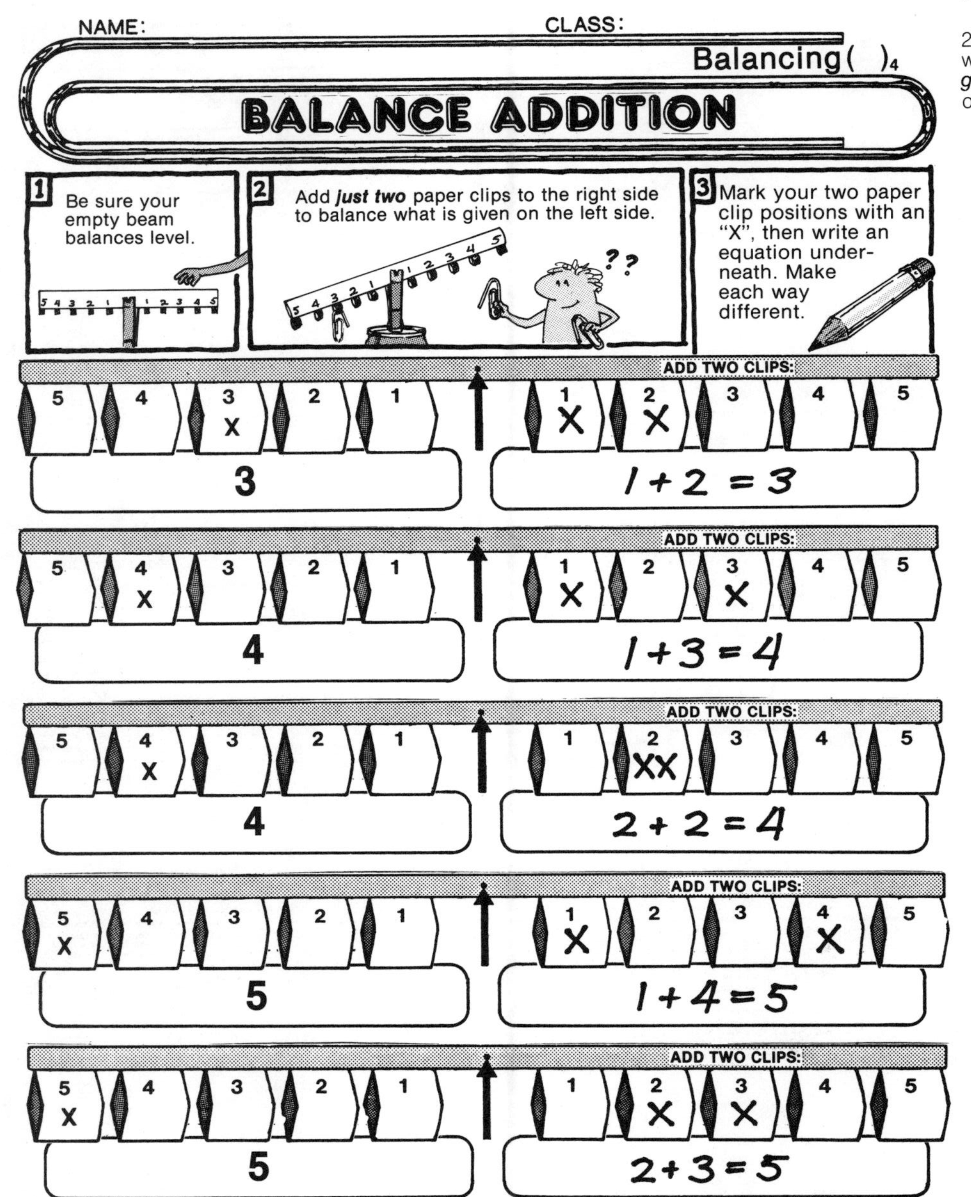

1. Throughout these exercises, be sure to emphasize the importance of checking *first* to see that the empty beam balances level, ***before*** adding paper clips. In the routine course of handling, the beams sometimes begin to tilt off balance, requiring the addition of an extra bit of tape to the higher (lighter) side.

2. Some students may not initially understand what to do. Explain that they should add the one ***given*** paper clip ***first***, then add two more to the other side to make the beam balance level.

Evaluation

Q: Draw ***two*** more X's to make this beam balance. Choose your positions carefully so you can write a ***different*** equation in each box.

A:

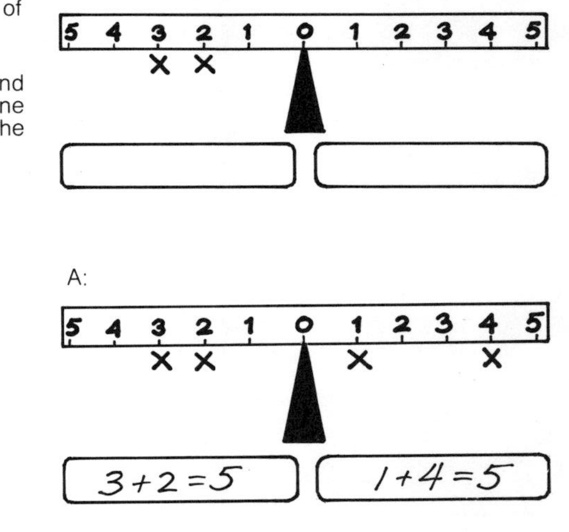

Materials

- ☐ A math balance.
- ☐ Paper clips of uniform size and weight.

(TO) understand that paper clips multiply to equal products on each arm of a *balanced* beam.

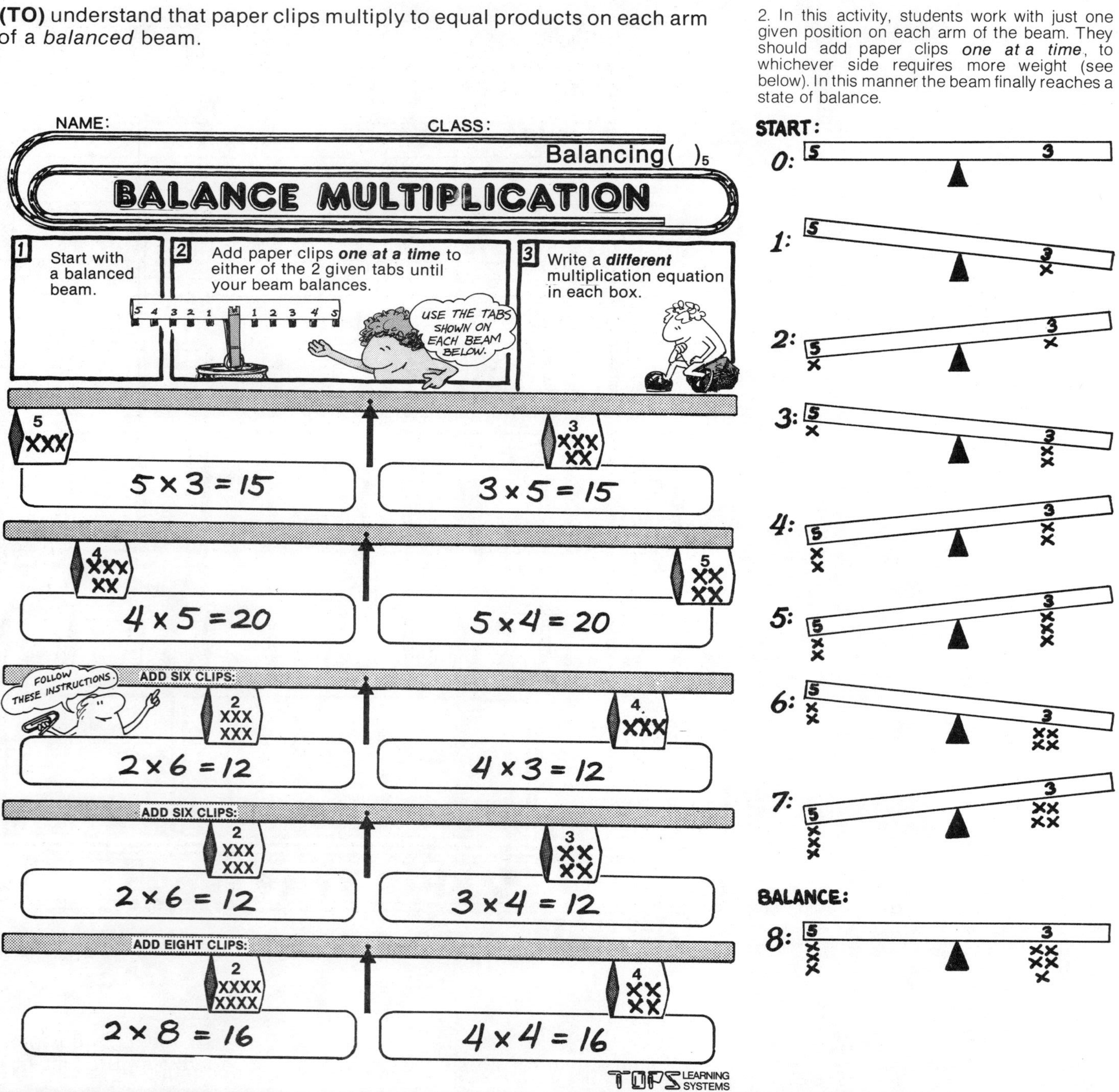

2. In this activity, students work with just one given position on each arm of the beam. They should add paper clips ***one at a time***, to whichever side requires more weight (see below). In this manner the beam finally reaches a state of balance.

Notice in this example that if the paper clips were added in bunches, it is quite possible that the first equal multiple (5x3 equals 3x5) would be passed over. In this case the balance wouldn't center again until the next higher multiple (5x6 equal 3x10). But with the addition of greater numbers of paper clips, the beam loses sensitivity, tilting less perceptibly with the addition of each new clip. That's why the directions ask students to add clips just ***one at a time***.

Evaluation

Q: Draw X's under positions 3 and 4 to make this beam balance. Write a multiplication equation in each box.

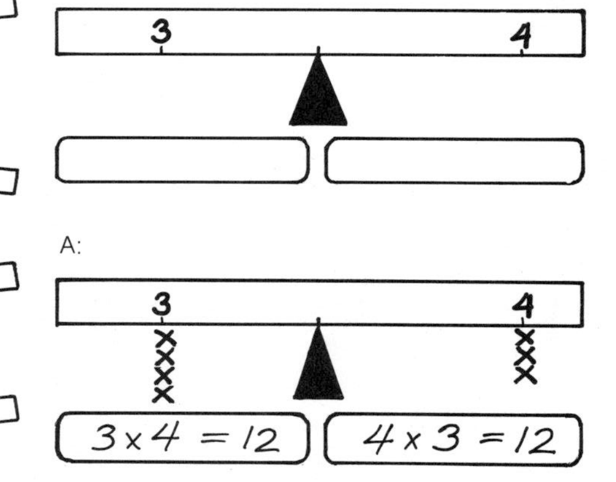

Materials

- ☐ A math balance.
- ☐ Paper clips.

(TO) gain further experience with balance beams and the mathematics of balancing.

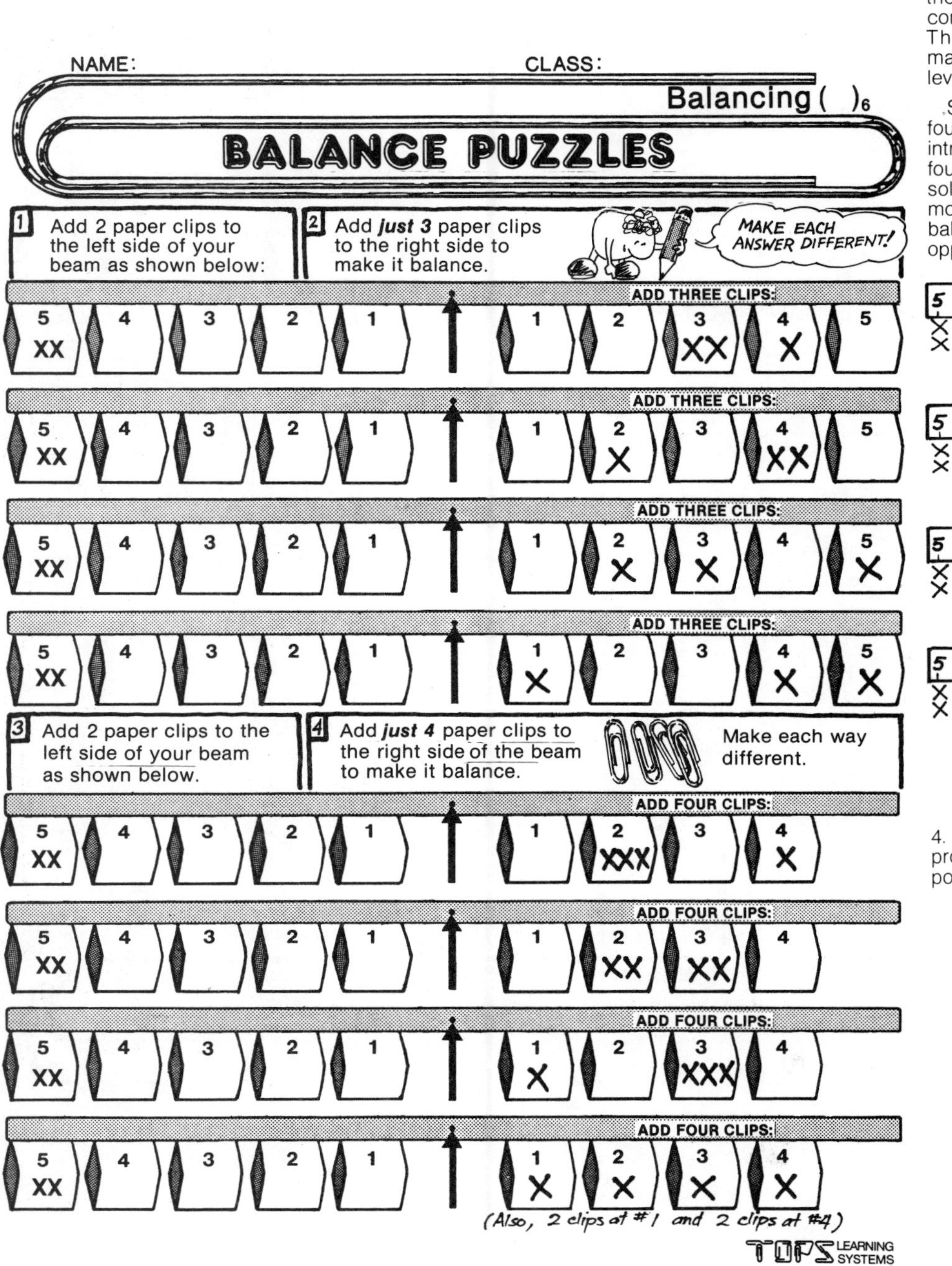

2. Once students understand the mathematics of the balance beam they tend to shift from physical activity (actually placing paper clips on the beam) to mental activity (writing number combinations that add up to the desired result). This is ideal. As a result of concrete manipulations, students advance to a higher level of mental abstraction.

Students who experience difficulty in finding all four balance combinations can be helped if you introduce the concept of symmetry. In the first four problems on the worksheet, notice how each solution is derived from the last by symmetry moves. Once the beam is placed in a state of balance, then clips that are moved in equal but opposite directions will maintain that balance.

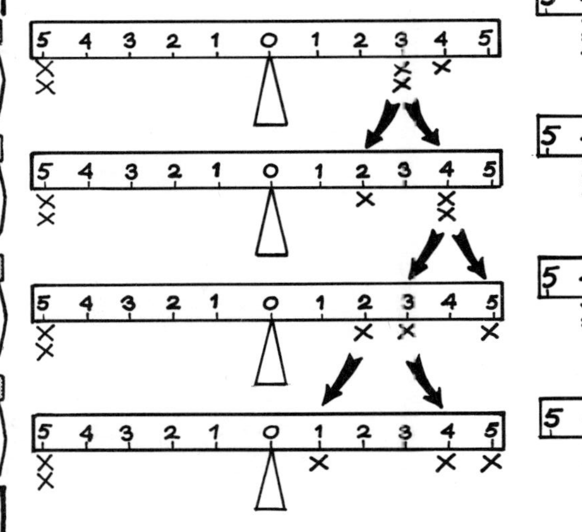

4. Notice that solutions to this last group of problems are restricted to the first four beam positions.

Evaluation

Q: Draw 4 different ways to make this beam balance by adding *three* paper clips to the right arm.

A:

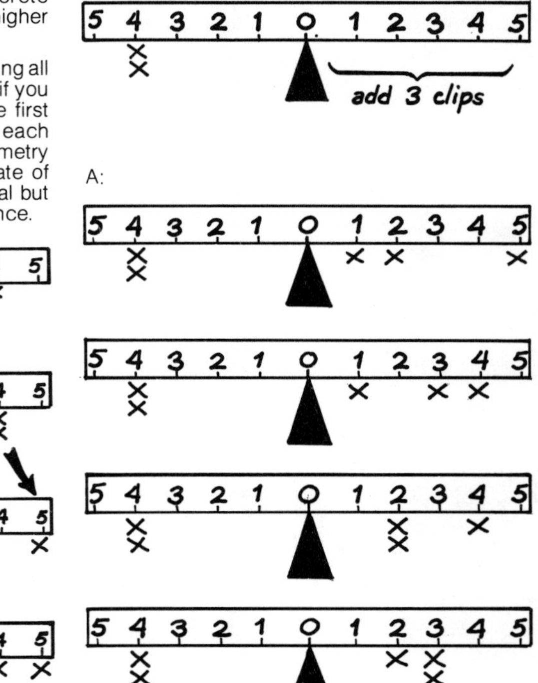

Materials

- ☐ A math balance
- ☐ Paper clips.

(TO) practice expressing complex balance conditions as mathematical equations.

1-2. Some students will likely complete this activity without using a math balance. Others may need to use the balance in order to solve the puzzles. In either case, the exercise establishes a strong connection between mathematics and the nature of balancing beams.

Each answer is written so the numbers correspond to the physical placement of the paper clips on the beam. When multiplication is involved, a pre-algebra product notation "X" is used with the *position* written first, followed by the *number* of paper clips at that position. Because these "rules" are followed naturally by most students, they are used here. Alternate forms, of course, are also acceptable, as long as the equations are mathematically correct.

Evaluation

Q: Write an equation in each box. Does this beam balance?

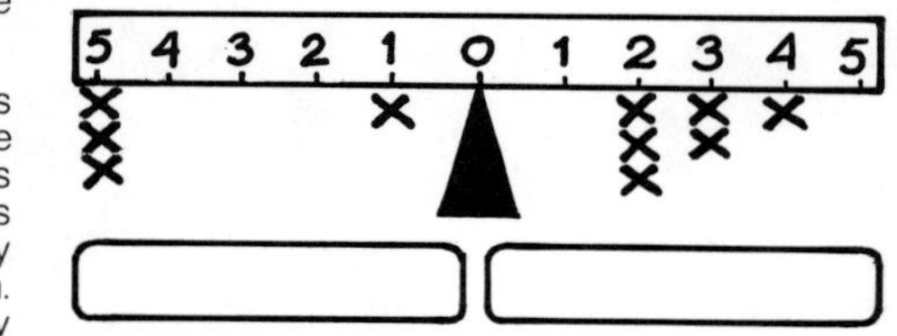

A: Yes.

$5 \times 3 + 1 = 16$ | $2 \times 3 + 3 \times 2 + 4 = 16$

Materials

- ☐ A math balance.
- ☐ Paper clips.

(TO) mathematically predict and then verify a state of balance or imbalance in a math beam.

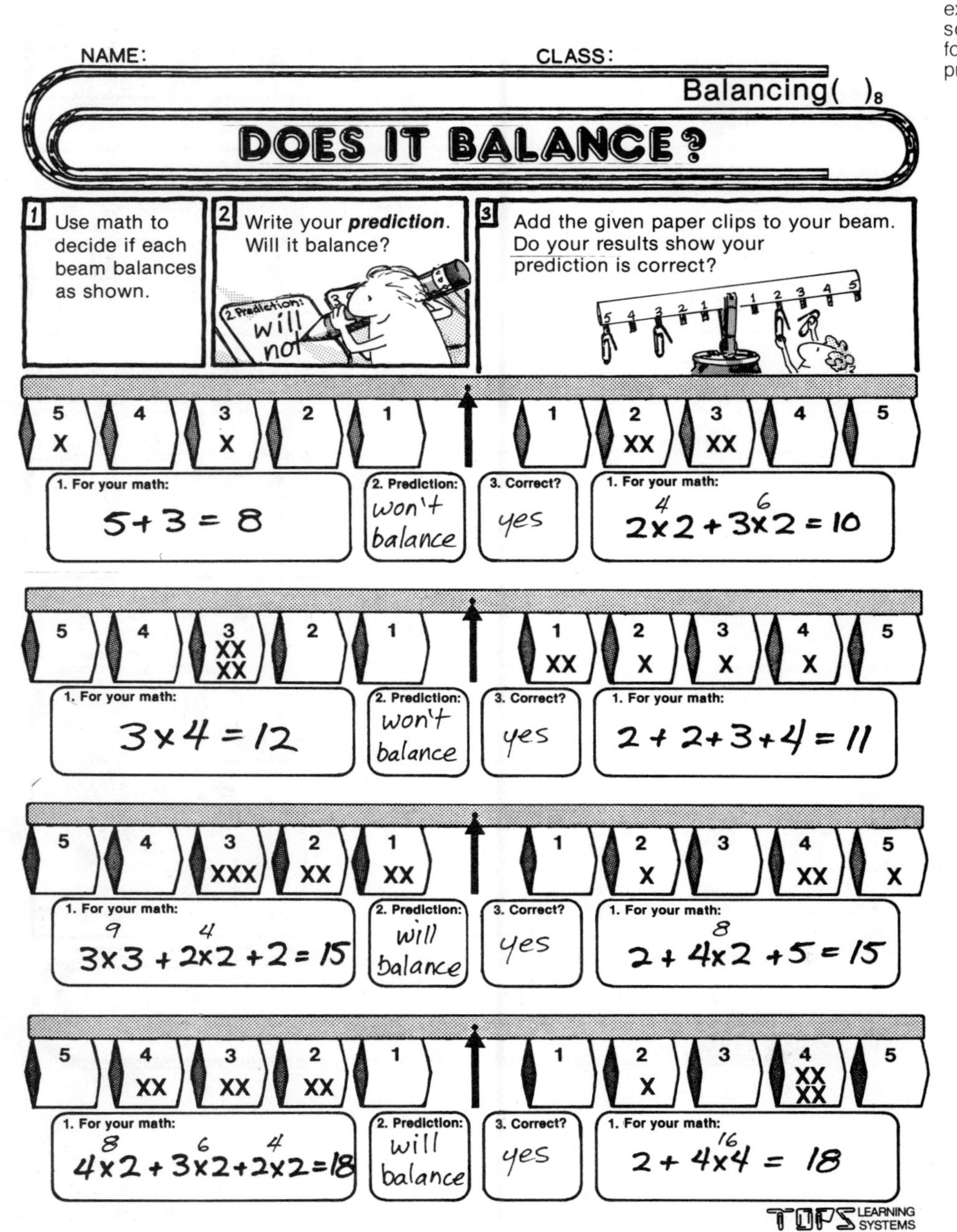

This activity requires students to ***predict*** a state of balance (or imbalance) based on their mathematical calculations. Then they ***verify*** by observation that their prediction is correct. This process of thoughtful prediction followed by experimental verification is fundamental to doing science. Insist, therefore, that your students follow these steps in sequence: (1) calculate (2) predict (3) verify.

Evaluation

Q: Find the beam that balances. Show your math.

a. 5 4 3 2 1 0 1 2 3 4 5

b. 5 4 3 2 1 0 1 2 3 4 5

c. 5 4 3 2 1 0 1 2 3 4 5

A: Beam "b" balances.

a. L: $5 + 3 + 1 = 9$
R: $2\times2 + 4\times2 = 12$

b. L: $5 + 2\times3 + 1 = 12$
R: $2 + 3\times2 + 4 = 12$

c. L: $4\times2 + 2\times3 = 14$
R: $2\times3 + 3\times2 + 4 = 16$

Materials

- ☐ A math balance.
- ☐ Paper clips.

(TO) mathematically predict and then verify the tilt of a math beam.

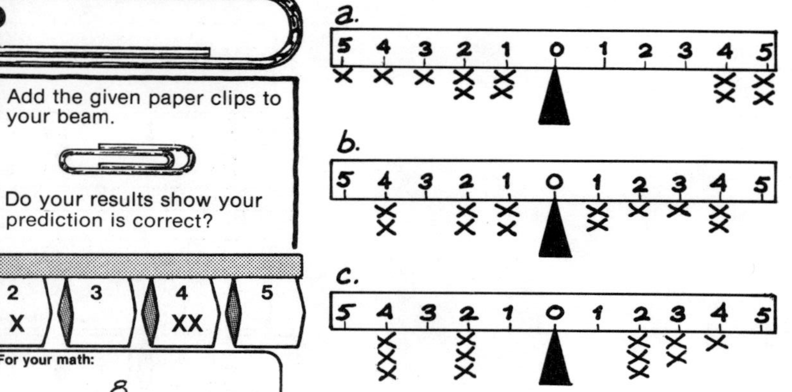

Beam 1 — Left: 5 (none), 4 X, 3 X, 2 XX, 1 (none) | Right: 1 (none), 2 X, 3 (none), 4 XX, 5 (none)

1. For your math: $4 + 3 + 2\times2 = 11$ | 2. Prediction: tilts left | 3. Correct? yes | 1. For your math: $2 + 4\times2 = 10$

Beam 2 — Left: 5 XXXX, 4 (none), 3 XXX, 2 (none), 1 (none) | Right: 1 X, 2 X, 3 (none), 4 XXXXXX, 5 (none)

1. For your math: $5\times4 + 3\times3 = 29$ | 2. Prediction: tilts left | 3. Correct? yes | 1. For your math: $1 + 2 + 4\times6 = 27$

Beam 3 — Left: 5 (none), 4 XXX, 3 (none), 2 X, 1 (none) | Right: 1 (none), 2 XXXXXX, 3 (none), 4 (none), 5 X

1. For your math: $4\times3 + 2 = 14$ | 2. Prediction: tilts right | 3. Correct? yes | 1. For your math: $2\times6 + 5 = 17$

Beam 4 — Left: 5 X, 4 X, 3 XX, 2 X, 1 X | Right: 1 (none), 2 XXXXX, 3 (none), 4 XX, 5 (none)

1. For your math: $5 + 4 + 3\times2 + 2 + 1 = 18$ | 2. Prediction: balances | 3. Correct? yes | 1. For your math: $2\times5 + 4\times2 = 18$

TOPS LEARNING SYSTEMS

As in activity 8, be sure your students follow these steps in sequence: (1) calculate (2) predict (3) verify.

Evaluation

Q: Find the beam that tilts ***left***. Show your math.

a. Beam positions 5 4 3 2 1 0 1 2 3 4 5 — Left: 5 X, 4 X, 3 X, 2 XX, 1 XX | Right: 4 XX, 5 XX

b. Beam positions 5 4 3 2 1 0 1 2 3 4 5 — Left: 4 XX, 2 XX, 1 XX | Right: 1 XX, 2 X, 3 X, 4 XX

c. Beam positions 5 4 3 2 1 0 1 2 3 4 5 — Left: 4 XXX, 2 XXX | Right: 2 XXX, 3 XX, 4 X

A: Beam "c" tilts left.

a. L: $5 + 4 + 3 + 2\times2 + 2 = 18$
R: $4\times2 + 5\times2 = 18$

b. L: $4\times2 + 2\times2 + 2 = 14$
R: $2 + 2 + 3 + 4\times2 = 15$

c. L: $4\times3 + 2\times3 = 18$
R: $2\times3 + 3\times2 + 4 = 16$

TEACHING NOTES 9

Extension

In this classroom demonstration, a student's weight is calculated by balancing on a two-by-four opposite another student of known weight. This "teeter-totter" system is completely analogous to a paper clip math balance: the numbered positions on the paper beam correspond to the ***distance*** that each student stands from the center pivot. The paper clips used on the beam correspond to the ***weight*** (in pounds or kilograms) of each student standing on the beam.

PREPARATION

Mark an 8 to 10 foot two-by-four into 16 equal subdivisions. This is best accomplished by cutting a piece of string equal to the length of the beam, then folding it into half lengths, quarter lengths, etc.

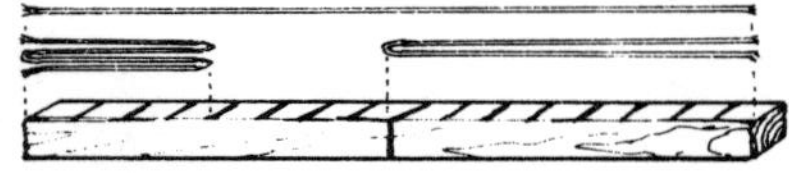

Use a block of wood or a brick as the pivot. If the beam will not balance at your center mark add a flat rock or other appropriate weight as a "rider"

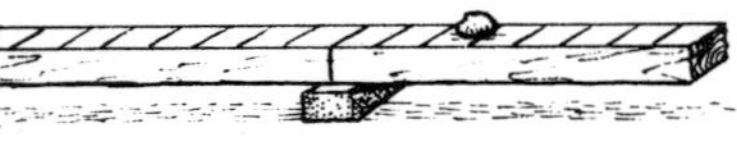

Obtain a bathroom scale in pounds or kilograms to weigh students and check experimental results.

PRESENTATION

Ask for two volunteers from your class. Weigh one of them on the bathroom scale. Calculate the weight of the other using balance beam mathematics. Caution: the two-by-four tends to be unstable when stood upon. Use "spotters" to help your volunteers balance safely.

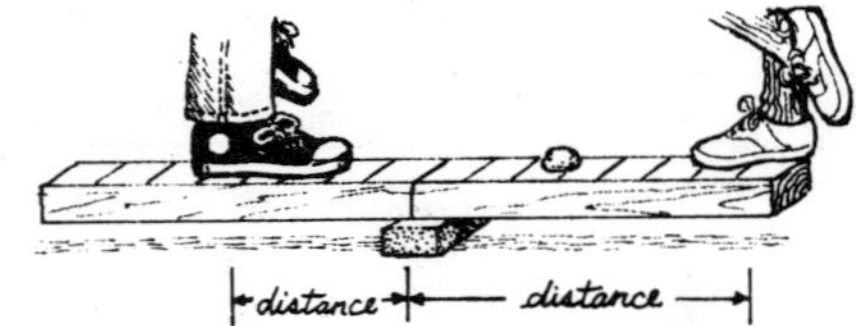

Materials

☐ A math balance.
☐ Paper clips.
☐ The optional extension activity requires extra materials. See above.

(TO) recognize that the mathematical properties of an *equal* arm balance can be extended to a balance with *unequal* arms.

NAME:

CLASS:

Balancing( )10

SHORT 'N' LONG ARM BALANCING

1 Pull the pin out of the center pivot.

2 Push it back through the mark above #1 to the ***left*** of zero.

3 Push this tab up ***Inside*** the beam, out of the way.

4 Put your off-center beam back on the clothespin.

Make it balance level by adding clay to the short arm.

5 Here is one way to make 5 paper clips balance on your short n' long arm balance beam.

Show 3 other ways to make 5 paper clips balance.

6 Your short n' long arm beam can still add and multiply if you renumber it. Begin at the new pivot and call it "0". Cross out the old numbers and write in the correct numbers on each tab.

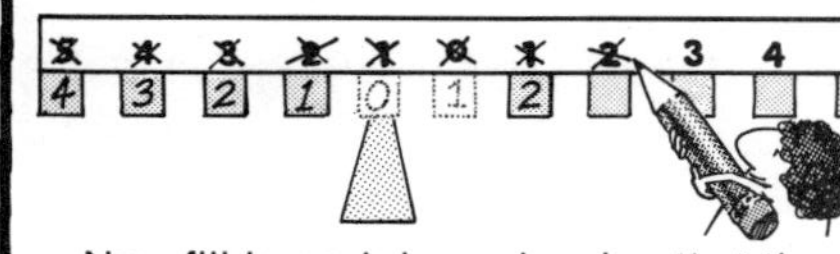

Now fill in each box, showing that these beams add and multiply correctly.

TOPS LEARNING SYSTEMS

2. A styrofoam cup makes a good "pin cushion", allowing the pin to be punched through the beam with relative ease. A clothespin will also serve.

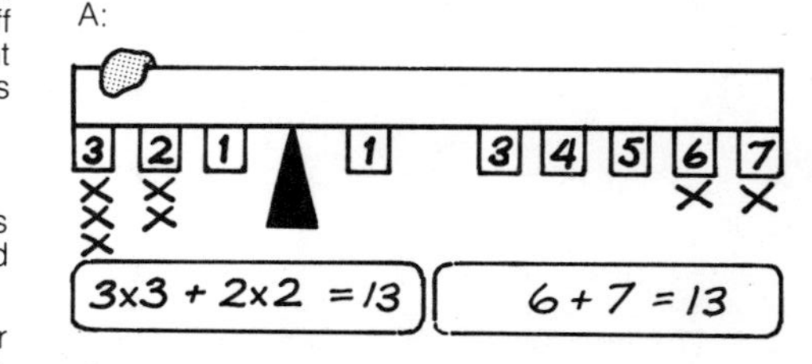

3. Since this is the last activity using math beams, the tab can just as well be snipped off with scissors. Your students, however, may want to preserve their balance beams intact, perhaps taking them home.

5. There are many ways to make 5 paper clips balance. Your students will probably find combinations different than these.

Because the numbers on the beam no longer correspond to actual distances from the pivot, solutions such as these are best found by actual trial and error experimentation on the long'n'short arm balance.

6. Once distances to the center pivot are properly written, students will discover that asymmetrical beams, like equal-arm beams, exhibit identical mathematical properties.

Evaluation

Q: Write the correct number in each tab. Then write an equation in each box to show that the beam balances.

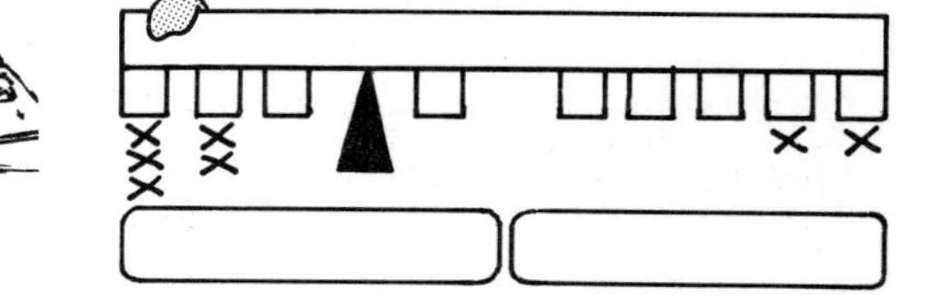

A:

Materials

- ☐ A math balance.
- ☐ Paper clips.
- ☐ A lump of modeling clay.
- ☐ A styrofoam cup is optional. See step 2 above.

(TO) fold a paper beam to use in a weighing balance.

BUILD A PAPER BEAM BALANCE (1)

1 Carefully trim this paper along the dotted lines.

2 Fold it in half ***exactly*** along the center line.

Keep the printing on the outside.

3 Fold ***up*** this half again so that all the edges meet.

FOLD UP BOTH EDGES TOGETHER.

FOLD

8 7 6 5 4 3 2 1 CENTER

4 Fold UP this quarter again so the edges meet.

This is your third fold.

5 Tape where shown to form a single closed strip.

6 Make pinholes through the 3 ***circled*** crosses on the other side.

7 Write your name here:

TOPS LEARNING SYSTEMS

TEACHING NOTES 11

Before your students build their own weighing balances in this activity and the next, construct one yourself. This will familiarize you with the directions, and provide a model for your students to follow.

1. Just like the math balances constructed in activities 1-2, this student worksheet also folds into an actual balance beam. These directions are similar enough to the ones before, that students should experience few difficulties in completing the steps successfully.

2. When folded together along the guideline, the top and bottom edges of the paper don't quite meet. These edges will "creep" together as the beam is doubled over in step 3 and again in step 4.

Some students may fold the paper inward and thereby cover the directions. The illustration directs students to fold the worksheet outward.

3. ***Both*** bottom edges of the paper must be folded ***up*** as illustrated, exposing step 4. Students who fold it down will be lead directly into step 5, likely skipping step 4 altogether.

4. Again, the paper must be folded ***up*** to lead into steps 5, 6 and 7.

6. Punch a pin through the 3 ***circled*** crosses only. The smaller crosses above each number should be left alone for now. They pertain to activity 20.

Before your students reach this step, you might demonstrate how to easily push a pin through the beam. A styrofoam cup serves as a convenient "pin-cushion". Tape helps protect sensitive fingers.

7. The completed weighing beam should look something like this.

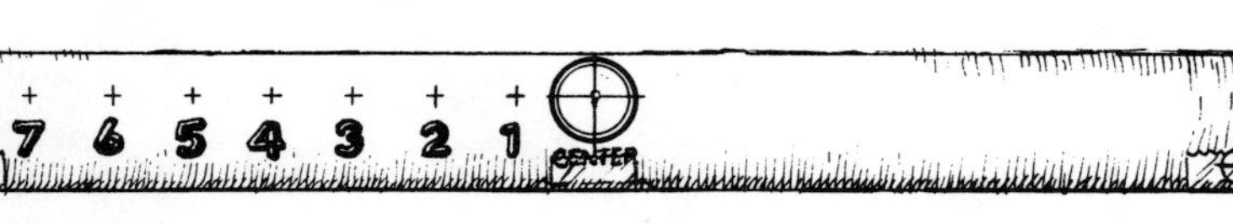

Evaluation

Is the paper well folded so the edges match more or less evenly? Are the three circled cross marks punched through right on center?

Materials

- ☐ A pair of scissors.
- ☐ Cellophane tape.
- ☐ A straight pin.
- ☐ A styrofoam cup.

(TO) complete construction of the weighing balance.

NAME: CLASS:

Balancing ( )12

BUILD A PAPER BEAM BALANCE (2)

1. Fold a 3 x 5 index card in half both ways. Cut the long fold to the center.

Cut to center

2. Fold up the tabs. Trim and tape.

a. Fold b. Trim c. Tape

3. Unbend a paper clip. Tape the small end to the outside of the folded-up card.

4. Cut into the opposite end about 1 cm (the width of a paper clip). Fold up and tape.

a. Cut b. Fold c. Tape

5. Repeat these steps to make a second weighing "pan".

6. Sharpen your pencil to a fine point. Use it to make the two ***end*** pinholes about as large as a pinhead.

DO ONLY THE TWO END HOLES.

Poke against the bottom of a styrofoam cup.

7. Poke 2 more unbent paper clips through the enlarged holes. Stick a pin through the center.

8 7 6 5 4 3 2 1

SMALL END

LARGE END

Stick a pin back through the center hole.

8. Bend the paper clip on each weighing pan forward, then loop the free end.

BEND DOWN

CLOSE LOOP

9. Balance your paper beam on a pin and taped clothespin as before. Hang a pan at each end, then center the beam with a clay rider.

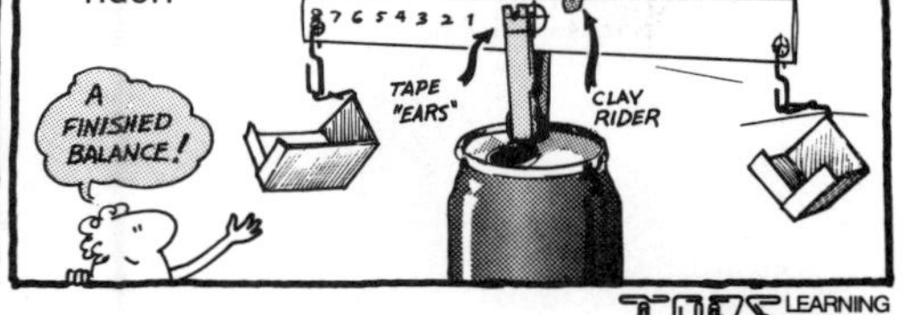

TOPS LEARNING SYSTEMS

6. This should not be attempted with a dull pencil. Students who do so will likely "dog-ear" the corners of the beam and perhaps tear a gaping hole in the top of the styrofoam cup. Use instead a ***very*** sharp pencil. Place the point on the pin hole, then rotate the pencil back and forth between your fingers like a drill. This will produce a clean, well defined hole. Care should be taken not to make the holes too large. Drilling it to the size of a pinhead allows an inserted paper clip to rotate freely, but restricts the hole from becoming so large that the paper clip wobbles from side to side in the hole.

7. In general, the beam becomes more stable as you place the pin higher and more unstable (but more sensitive) as you place the pin lower.

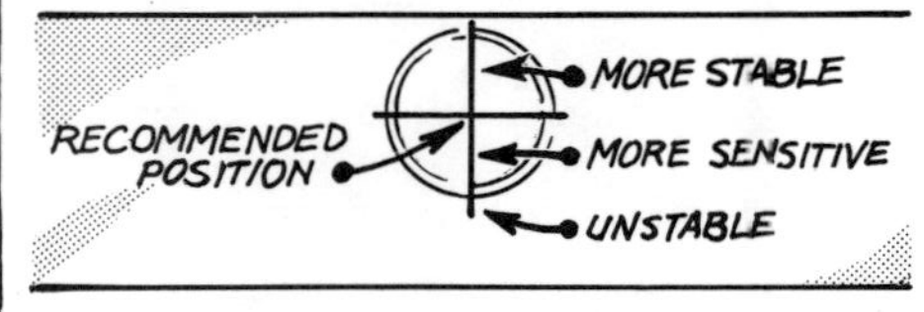

The intersection of the cross marks is only a recommended compromise between stability and sensitivity. Make adjustments up or down the vertical line as necessary.

8. The paper clips should be bent forward to right angles. This insures that the pans will hang at level positions from the beam. The loops at the ends of the paper clips keep the pans from inadvertently falling off the balance as students weigh things.

9. The clothespin support system used here is the same as the one used to the hold the math balance. See teaching notes 2, steps 1-3.

Depending on how much tape you used to put each weighing pan together, one likely weighs a little more than the other. Whenever these pans are interchanged, you'll need to shift the rider right or left to restore the overall balance. Clay is used instead of tape to facilitate frequent recentering.

Evaluation

Does the balance return to a centered level position each time the beam is tipped off center? Do the paper clips at each end rotate freely, so the pans remain level as the beam is tilted from side to side?

Materials

- ☐ Two index cards.
- ☐ A pair of scissors.
- ☐ Tape.
- ☐ Paper clips.
- ☐ A lump of modeling clay.
- ☐ A styrofoam cup.
- ☐ A suitable base for the beam. Use the same support system as you did for the math balance in activity 2. This will consist of a clothespin, straight pin and soda can or bottle.

(TO) make weight comparisons on a balance beam and thereby generate simple mathematical relationships.

NAME: CLASS:

Balancing( )13

SEEDS & PAPER CLIPS

1 Center your beam. Add seeds and paper clips to fill in each box.

10 paper clips = 12 pinto beans

10 pinto beans = 38 popcorns

10 popcorns = 24 lentils

10 lentils = 31 rice

2 Fill in each box again. Don't weigh again. Just divide by 10!

1 paper clip = 1.2 pinto beans

1 pinto bean = 3.8 popcorns

1 popcorn = 2.4 lentils

1 lentil = 3.1 rice

3 Use math to solve the equations in each of these boxes . . .

USE THE INFORMATION YOU FILLED IN ABOVE!

1 paper clip = ? popcorns	1 pinto bean = ? lentils	1 popcorn = ? rice
(a) Math Calculations: 1.2 × 3.8 = 96 + 36 = 4.56	(b) Math Calculations: 3.8 × 2.4 = 152 + 76 = 9.12	(c) Math Calculations: 2.4 × 3.1 = 24 + 72 = 7.44

NOTE: Answers will vary due to size variations among seeds.

. . . then check each answer on your balance.

Write the actual number in each box.

(a) Balance Check:	(b) Balance Check:	(c) Balance Check:
about 5	about 10	about 8

TOPS LEARNING SYSTEMS

1. If the beam does not balance exactly level, students should compensate by sliding the clay rider to a new position on the beam.

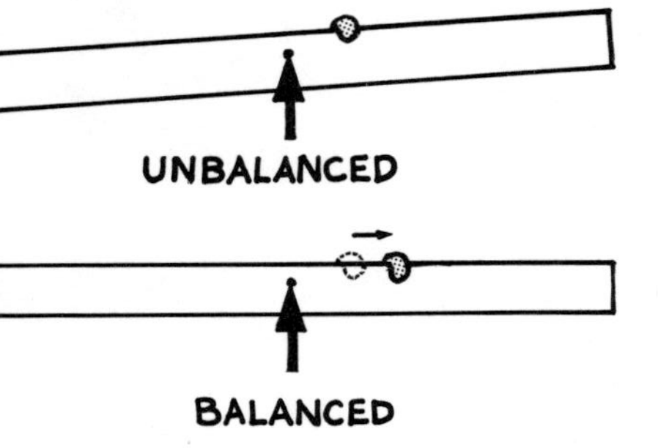

There is considerable size variation among seeds of the same kind, particularly among pinto beans and popcorn seeds. Answers throughout this activity, therefore, may be expected to vary by one or two seeds.

2. To fill in this second series of equations, students simply need to divide by 10 (move the decimal 1 place to the left).

3. You must *multiply* to find each answer. A simple dollar and cents analogy illustrates this operation. Suppose you know that:

1 dollar = 4 quarters
1 quarter = 5 nickels
1 nickel = 5 pennies

Then you can solve these other equations simply by multiplying.

1 dollar = ? nickels
4 x 5

1 quarter = ? pennies
5 x 5

When doing the balance check, students should select average-size seeds to weigh. Agreement with the math calculations will only be approximate due to variations in seed size.

TEACHING NOTES 13

Evaluation

Q: Ten pennies weigh 52 paper clips, and 10 paper clips weigh 150 staples. How many staples does a penny weigh?

A: Dividing by 10 we find:

1 penny = 5.2 paper clips
1 paper clip = 15 staples

Thus,

1 penny = 5.2 x 15 staples = 78 staples

Materials

☐ The paper beam balance constructed in activities 11 and 12.
☐ Paper clips.
☐ Four kinds of seeds: pinto beans, popcorn, lentils, and long-grain white rice.

(TO) weigh common classroom objects using a paper clip weight standard.

1. It is unlikely that the objects will weigh an exact number of whole paper clips. Encourage students to use a plus or minus for items that weigh a little more or less than a whole number of clips. In the next activity students will have the opportunity to weigh these same items to the nearest *tenth* of a paper clip.

These answers are based on medium-sized #1 paper clips. Remember to circulate only one brand in your classroom to maintain uniform weight.

2. Anything that fits into the pan is appropriate to weigh. Other suitable objects might include a nail, a short pencil, an eraser, a bottle cap, a piece of chalk, a cork or a button.

3. Encourage students to puzzle out this problem on their own. They will probably express their answers as fractions. Converting to the decimal equivalent gives good division practice.

Evaluation

A sheet of ditto paper weighs about 9 paper clips. Cut several sheets to size so they weight different numbers of ***whole*** paper clips, say 4, 5, and 9. Fold each as small as possible and label them A, B, and C respectively. Present them to your students.

Q: Weigh papers A, B, and C to the nearest ***whole*** paper clip on your paper beam balance.

A: A = 4 paper clips, B = 5 paper clips, C = 9 paper clips.

Materials

- ☐ A paper beam balance.
- ☐ Paper clips of uniform size and weight.
- ☐ Objects to weigh: a penny, sheet of paper, washer, and index card. Have other small common objects available as well. See step 2.

(TO) develop and use a system of weight measure accurate to a tenth of a paper clip.

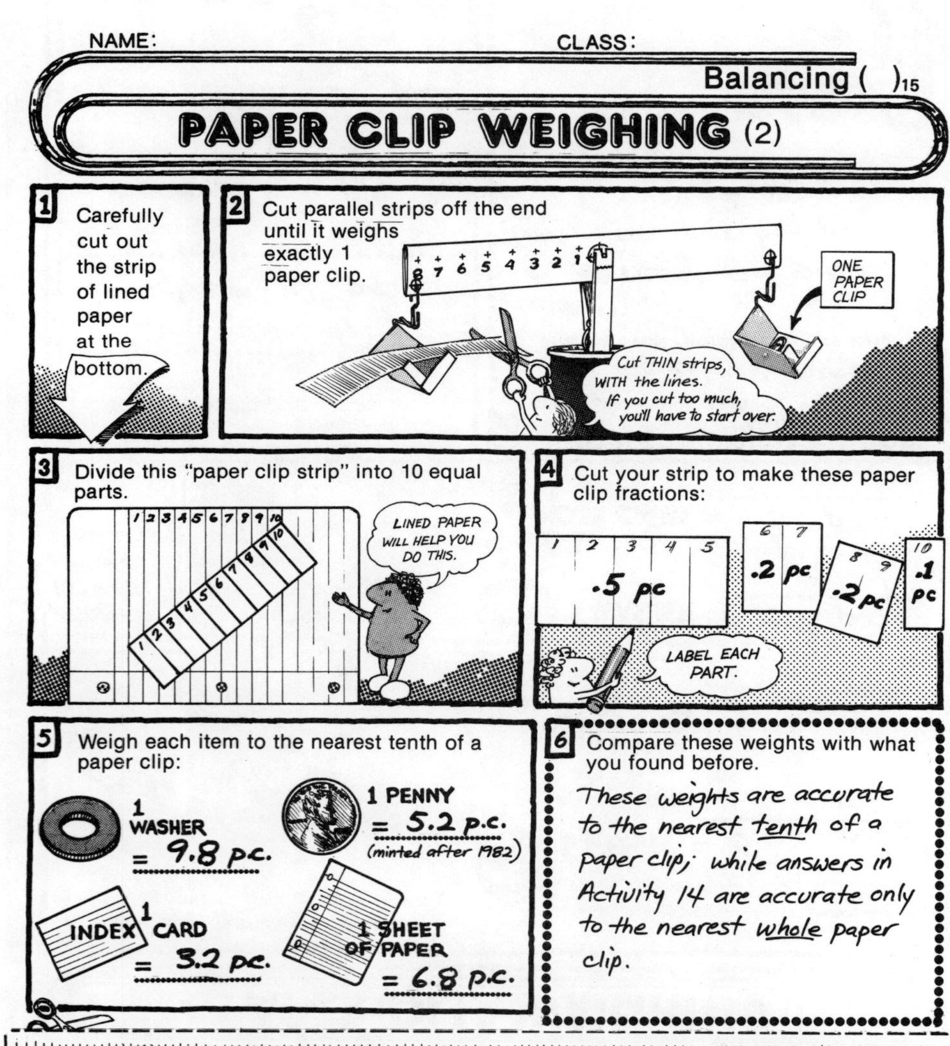

2. You may wish to run off a few extra copies of this worksheet so that students who cut their strip too short can try again using a new one.

3. Students often confuse the number of lines they should draw (9) with the number of spaces (10). Numbering the ***spaces*** as shown reduces this confusion.

4. Student answers may vary from the ones given here because they possibly use a different brand of paper clips, a different size of washer, etc. Once students have made their weight determinations, however, the ***range*** of answers for your particular class should not vary beyond several tenths of a paper clip.

Extension

Do all pennies weigh the same? Compare pennies minted in different years on your balance. Write a report.

In 1982 the U.S. government began minting a lighter penny, one that contained much more aluminum and much less copper. During this same year, heavier pennies were also produced according to the old formula. So your students should discover weight differences of about 1 paper clip for pennies minted during this year.

Copper pennies minted ***before*** 1982 are uniformly heavier, varying by only a few tenths of a paper clip. Likewise, aluminum pennies minted ***after*** 1982 are uniformly lighter, again varying only by a few tenths of a paper clip.

Evaluation

Cut a sheet of paper into 3 different sizes. Fold each as small as possible and label them X, Y, and Z respectively. Tape them with clear cellophane tape, if you wish, and present them to your students.

Q: Weigh papers X, Y, and Z to the nearest *tenth* of a paper clip on your balance.

A: Determine the "official" weight of each folded paper to the nearest tenth. Student answers should not vary from your own official weight by more than a tenth of a paper clip.

Materials

- ☐ A paper beam balance.
- ☐ Paper clips.
- ☐ A pair of scissors.
- ☐ A sheet of notebook paper.
- ☐ Objects to weigh: a penny, sheet of paper, washer and index card.

(TO) weigh light things using a "paper square" weighing standard.

NAME: CLASS:

Balancing( )16

PAPER SQUARE WEIGHTS

1. Cut out each group of squares along the **dotted lines** only.

DON'T cut the SOLID lines.

2. Weigh each item in these "paper squares". If you need to add less than 1 square, cut one in halves or quarters.

ITEM	WEIGHT
1 pinto bean	8 paper squares
½ index card	15½ p.s.
10 popcorns	25 p.s.
10 rice grains	3 p.s.
1 paper clip	10 p.s.

3. Weigh a penny in "paper squares". Use ***only*** your own 28 squares and some clay. Tell how you did this.

First weigh the penny in squares and clay. Then weigh the clay alone.

1 penny = 28 sqs. + clay

clay = 25 sqs.

1 penny = 28 sqs. + 25 sqs.

= 53 sqs.

PENNY

PAPER SQUARES and CLAY

How many paper squares does the clay weigh?

4. SAVE YOUR SQUARES

PAPER SQUARES
CUT THE DOTTED LINES ONLY:

TOPS LEARNING SYSTEMS

1. Students should cut only ***groups*** of squares along the ***dotted*** lines. They should leave the individual squares, separated by solid lines, intact.

Remind students to save these squares for use in the next activity.

2. Because of size variations, the weights of the seeds may vary by several paper squares. If your students consistently get answers that are more (or less) than these, it is probably because you duplicated the squares using a heavier (or lighter) grade of paper.

3. Some students may pool their paper squares and thereby accumulate enough to weigh the penny directly. This is a good way to double-check the accuracy of their work, but the original problem still remains; how to weigh the penny using *only* 28 squares plus clay.

The worksheet answer is based on pennies minted after 1982. If your students select pre-1982 pennies, minted with more copper, they'll weigh more, perhaps 60 squares each.

Another way to solve this problem is to put enough clay in one pan to just counterbalance a penny in the other. Then divide the clay and weigh each part in paper squares. The total of course, will equal the weight of the penny.

Evaluation

Q: Find the weight of a nickel in paper squares. Use only your own 28 squares and some clay.

A: About 98 squares.

Materials

- ☐ A paper beam balance.
- ☐ An index card.
- ☐ A paper clip.
- ☐ Three kinds of seeds: pinto beans, popcorn, and long-grain white rice.
- ☐ A lump of modeling clay.

(TO) find the weight, in paper squares, of different kinds of seeds. To predict, among groups of seeds, which weighs more.

NAME: CLASS:

Balancing ( )17

HEAVY 'N' LIGHT WEIGHTS

1 Weigh these seeds in paper squares.

10 rice grains = 3 paper squares

10 lentils = 10 p.s.

10 popcorns = 25 p.s.

2 Divide by 10 to find the weight of one seed.

1 rice grain = 0.3 paper squares

1 lentil = 1.0 p.s.

1 popcorn = 2.5 p.s.

3 Predict which side will have more weight by filling in the table.

	LEFT PAN:	RIGHT PAN:	GREATER WEIGHT ?
a.	8 rice grains	2 lentils	rice
b.	12 rice grains	3 popcorns	popcorn
c.	8 lentils	2 popcorns	lentils

Show your math here...

a. rice: 0.3 × 8 = 2.4; lentils: 1.0 × 2 = 2.0

b. rice: 0.3 × 12 = 3.6; popcorn: 2.5 × 3 = 7.5

c. lentils: 1.0 × 8 = 8.0; popcorn: 2.5 × 2 = 5.0

4 Check each prediction on your balance.

a. correct?	b. correct?	c. correct?
yes	yes	yes

If you were wrong, try to find out why!

TOPS LEARNING SYSTEMS

2. To divide by 10, students need only move the decimal point 1 place to the left.

3. Be sure that students complete step 3 before proceeding to step 4. This insures that they complete each prediction before testing its validity.

4. Weight differences between the left and right pan are probably too large to be reversed by size variations among seeds. If students predict incorrectly, look for math errors or poor weighing technique. One common error is that students forget to center their beams before weighing.

Evaluation

Q: Ten rice grains weigh about 3 paper squares. One paper clip weighs about 10 paper squares. Calculate which has the greatest weight, 60 rice grains or 2 paper clips?

A: 10 rice grains = 3 paper squares, thus 60 rice grains = 18 paper squares

1 paper clip = 10 paper squares, thus 2 paper clips = 20 paper squares.

Two paper clips, therefore, weigh more than 60 rice grains.

Materials

- ☐ A paper beam balance.
- ☐ Three kinds of seeds: rice, lentils, and popcorn.
- ☐ Paper square weights from activity 16.

(TO) reconcile mathematically the weight ratios of paper rectangles with their area ratios.

NAME: CLASS:

Balancing( )18

SQUARES AND RECTANGLES

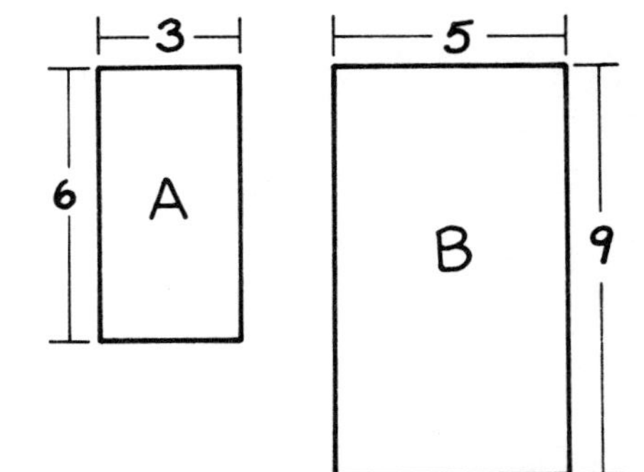

1 Carefully cut out the ruler ***plus*** each square and rectangle on the second sheet of paper.

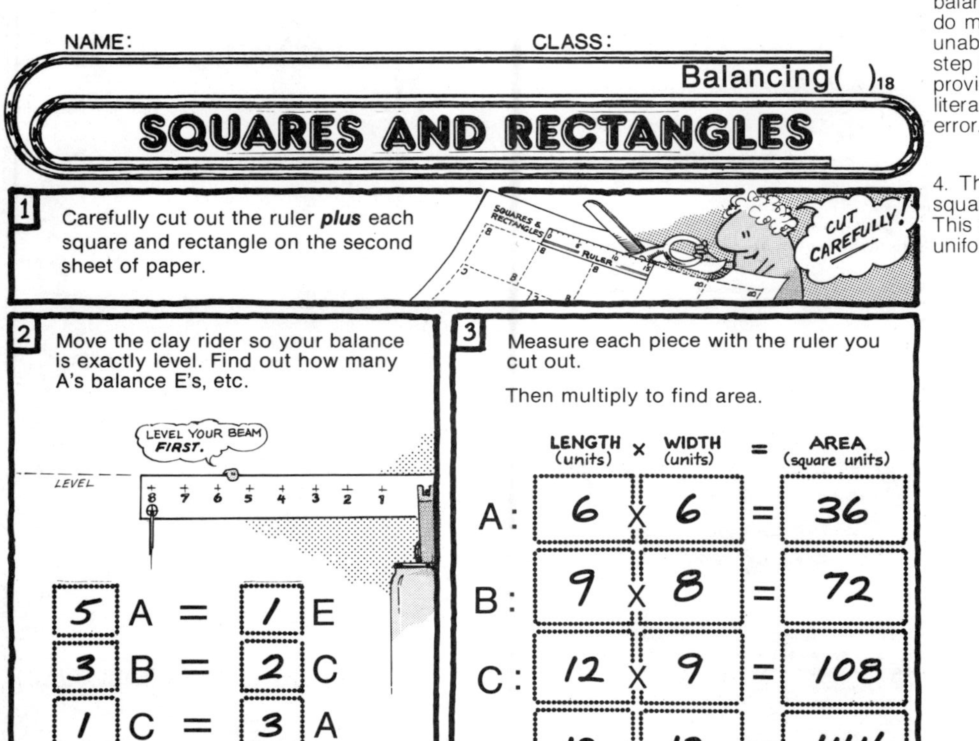

2 Move the clay rider so your balance is exactly level. Find out how many A's balance E's, etc.

5 A = 1 E
3 B = 2 C
1 C = 3 A
1 D = 2 B
2 E = 5 B

3 Measure each piece with the ruler you cut out.

Then multiply to find area.

	LENGTH (units)	×	WIDTH (units)	=	AREA (square units)
A:	6	×	6	=	36
B:	9	×	8	=	72
C:	12	×	9	=	108
D:	12	×	12	=	144
E:	15	×	12	=	180

4 Show that your ***areas*** in step 3 agree with your weight equations in step 2.

A	E	B	C	C	A	D	B	E	B
36	180	72	108	108	36	144	72	180	72
×5	×1	×3	×2	×1	×3	×1	×2	×2	×5
180	180	216	216	108	108	144	144	360	360

TOPS LEARNING SYSTEMS

2. Students who remember to center their beams before weighing the squares and rectangles should experience little difficulty in differentiating between combinations that *truly* balance dead level, and ones that *almost* balance, yet tilt perceptibly off center. Those who do make errors will find in step 4 that they are unable to reconcile the ***weight*** proportions of step 2 with the ***area*** proportions of step 3. Don't provide assistance too quickly. Part of scientific literacy is learning to back-track and isolate error.

4. These calculations show that the area of each square or rectangle is proportional to its weight. This is true for any material, like paper, that has uniform thickness.

Evaluation

Q: Paper A weighs 2 paper clips. What is the weight of paper B?

A (3 × 6) B (5 × 9)

A: Five paper clips. Weight is proportional to area. Since A fits into B 2½ times, B is 2½ times heavier.

Materials

☐ A paper beam balance.
☐ A pair of scissors.

(TO) "count" an unkown quantity of paper clips by comparing their weight to a known number of paper clips on a balance.

NAME: CLASS:

Balancing ( )19

EDUCATED GUESS

1 Start with 2 sheets of paper that are exactly the same size and weight.

2 Ask a friend to hide from 1 to 10 paper clips in one of the papers while you look away.

3 Use your balance to guess how many paper clips are wrapped inside.

My guess was ☐ correct ☐ wrong

4 Repeat this experiment until you guess the right amount at least 3 times in a row. Tell how you did it.

Center your balance. Place the wrapped-up clips in one pan and the unused paper (folded up) in the other. Add clips to the lighter side until your beam balances level. The number of clips you add equals the amount wrapped in the paper.

5 As a bank teller, your job is to count pennies to wrap in 50¢ rolls.

Tell how you would use a balance to make your job easier.

Center your balance and place 50 pennies in one of the pans. To "count" 50 more pennies, simply add them to the other pan until the beam returns to a centered level position.

TOPS LEARNING SYSTEMS

3. Some students may forget to add a sheet of paper to counter-balance the one wrapping the clips. They will erroneously predict a number of clips that is too large.

The best way to guess the number of clips is to add them to the pan one at a time until the beam balances Some students will likely follow a different "hit or miss" strategy, wrapping up the number of clips they think is correct, then adding them all at once to the balance. This way is less efficient, but should eventually lead to a correct guess.

4. This hide-and-guess game is popular with younger students. Some may wish to repeat the procedure more than 3 times.

5. In 1982 the U.S. government changed its formula for making pennies, replacing most of the copper with aluminum. As a result, post-1982 pennies weigh about 1 paper clip less than pre-1982 pennies. This count-by-weighing method only works when the pennies have uniform weight. They can be older or newer than 1982, but not a mix.

An easier way to count pennies is to stack them in a 50¢ pile, then form other piles of identical height. This stacking method takes advantage of a penny's uniform thickness.

Evaluation

Q: A piggy bank is full of dimes. How could you find out how much money was inside without breaking it open?

A: Counterbalance the full piggy bank against an identical empty bank and a known number of dimes.

Materials

☐ A paper beam balance.
☐ Sheets of paper with equal weight.
☐ Paper clips.

(TO) graph how the force required to balance an off-center beam increases as it is applied closer to the pivot.

NAME: CLASS:

Balancing( )20

MOUNTAIN OF PAPER CLIPS

1 Punch pinholes through all 7 crossmarks along your beam. It's easy if you lay the beam on a styrofoam cup.

TAKE OFF BOTH WEIGHING PANS.

Protect your finger with tape.

FOAM CUP

2 Hang a paper clip "hook" on the left side. Slide a regular clip over the right end so your beam balances level on its center pivot.

LEVEL

UNBEND TO FORM A HOOK

SLIDE TO BALANCE

3 Fill in the data table: Find the number of paper clips you need to balance the beam at each new pivot position. Then graph your results in a smooth line.

A. Fill in table.

B. Draw smooth graph line.

HOW MANY CLIPS AT #1?

PIVOT #1

LEVEL

DATA TABLE:

Number of clips to balance	Pivot Position
0	0
1+	1
3+	2
6	3
10	4
17	5
31	6
69	7

PIVOT POSITION

NUMBER OF CLIPS TO BALANCE

4 Could a "mountain" of paper clips hanging ***directly under*** pivot #8 raise the beam to a level position? Your graph is telling you the answer.

The graph line approaches, but does not cross, the heavy line at pivot position 8. At that point, even an infinite "mountain" of paper clips has insufficient weight to raise the beam to a level position.

TOPS LEARNING SYSTEMS

3. A whole number of paper clips will not always balance the beam at each position. Students should use a plus or minus to indicate when slightly more or less weight is required.

At pivot position #7, the paper clip hook may not be large enough to hold all the required clips (about 69). Moreover, the paper clip cluster may begin to touch the support can or bottle. These problems are solved by fashioning a ***double*** paper clip hook that holds ***more*** clips ***closer*** to the beam. Your students may discover other engineering solutions as well.

Students unfamiliar with graphing may require some assistance in plotting ordered pairs from the data table. To locate the "number of clips to balance" on the horizontal x-axis, students will need to know how to estimate between 1 and 10 on the number line. The following blackboard exercise will prepare your students.

*Locate these numbers on the number line: **5, 7, 2+, 15-, 11, 18***

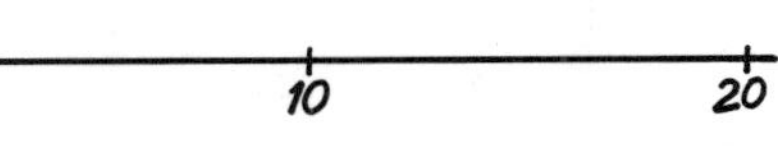

4. This rather abstract notion can be converted to concrete experience in the following manner: remove the beam from the balance and punch it through with a pin on the line directly above position 8. Suspend the pin from above with 3 additional paper clips. Students will easily observe that *no* amount of pulling force can raise the beam to a level position.

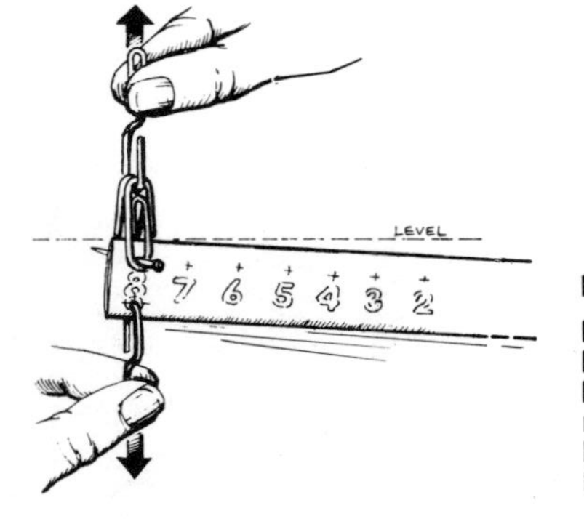

It is interesting to note that when the pivot is shifted very slightly to the right, say only a millimeter to position 7.9, the beam may be pulled to a level position with relative ease.

Evaluation

Q: Weights A and B are tied to identical wooden beams and balanced from ropes. Which weight is heavier? How do you know?

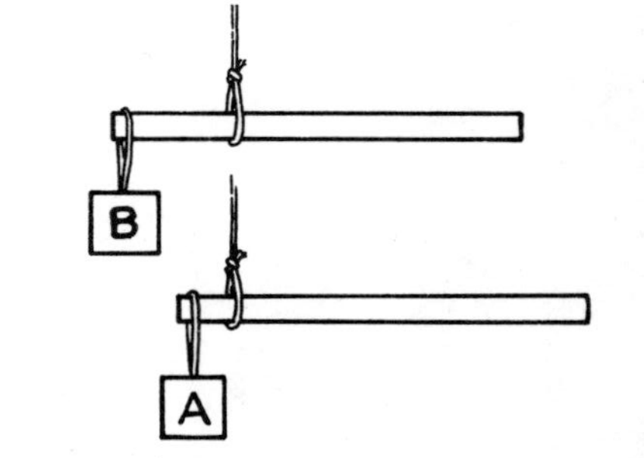

A: Weight A is heavier because it supports the weight of the beam at a position that is closer to the pivot.

Materials

☐ A paper beam balance.
☐ A styrofoam cup.
☐ A box with at least 75 paper clips. To keep the number of required paper clips within reasonable limits, this activity is best done on an individual basis or as a class demonstration.

HOW TO REMOVE WORKSHEETS

Perforated worksheets don't always work like they are supposed to.

This book is designed not only to make your science lessons run smoothly, but to make the worksheets pull out smoothly as well. Our pages are "perfect bound" in the same manner as single sheets of stationary are attached to a writing pad. You can remove worksheets from this book just like pulling sheets off a pad — well, almost.

We didn't want our book to shed leaves like a tree. So we ordered the perfect binding very strong. To remove the worksheets cleanly and quickly, be sure to follow one of these two special procedures.

One-at-a-time:

Start from the *back* of the book. *Pull* the top sheet off as illustrated. Don't tear. Proceed to the next until you remove all the worksheets.

The top sheet will probably be glued most securely to the binding. Sheets underneath should pull off more easily. Do not attempt to remove pages from the middle of the book *first.* This is often difficult to do (even on a scratch pad).

Radical Surgery:

Place a sharp knife on this very page with the edge facing the binding. Close the book, and pull the knife through the binding to cleanly remove all the worksheets. Strip off the back cover and separate each page.

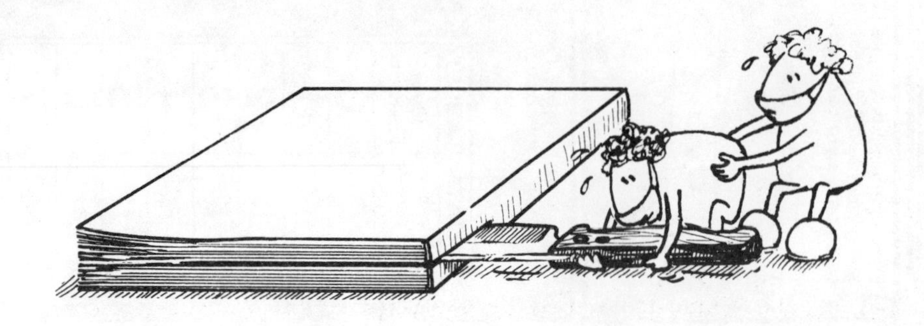

REPRODUCIBLE STUDENT ACTIVITY SHEETS

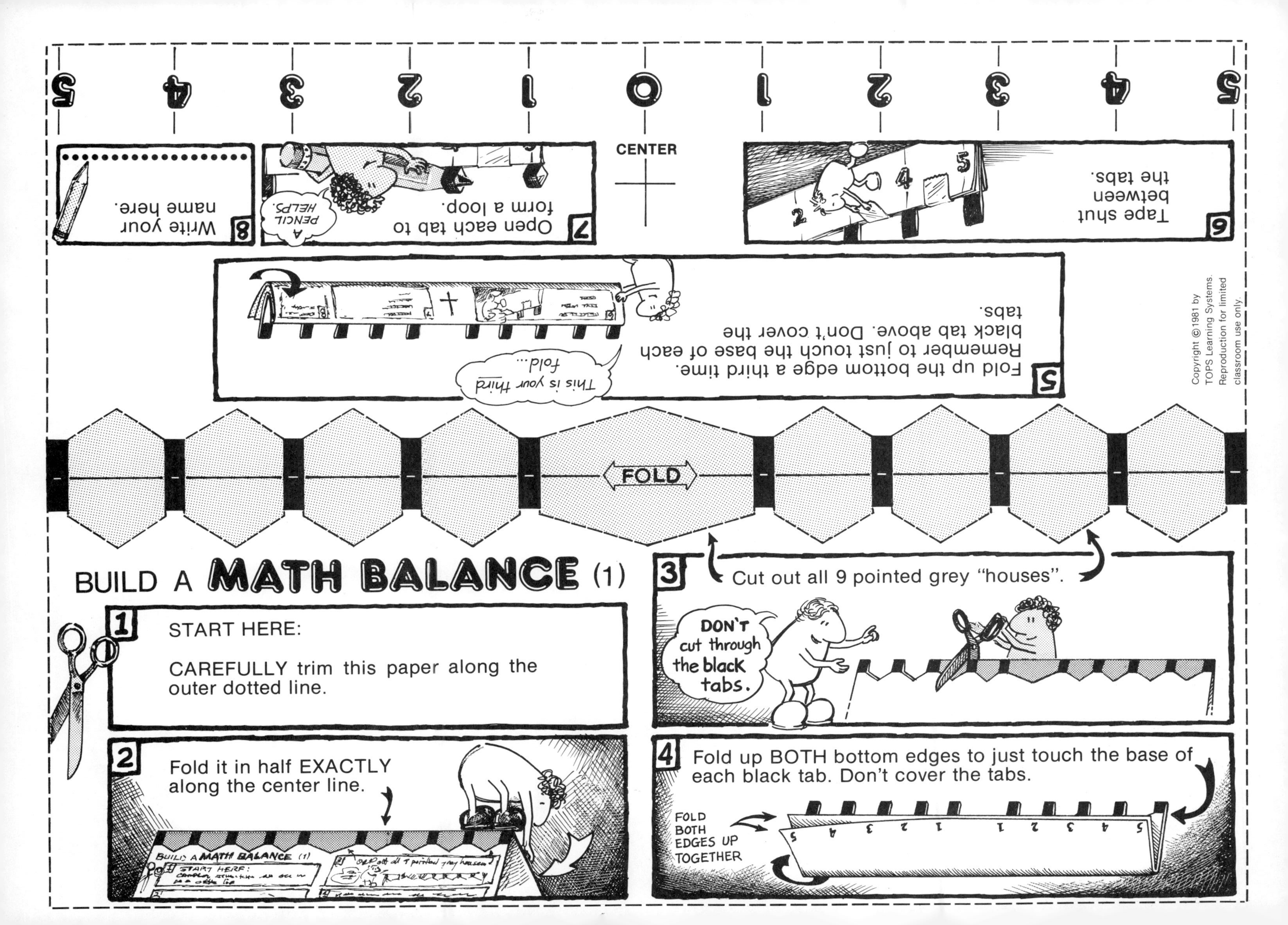

BUILD A MATH BALANCE (1)
1 START HERE:
CAREFULLY trim this paper along the outer dotted line.
2 Fold it in half EXACTLY along the center line.
3 Cut out all 9 pointed grey "houses".
DON'T cut through the black tabs.
4 Fold up BOTH bottom edges to just touch the base of each black tab. Don't cover the tabs.
FOLD BOTH EDGES UP TOGETHER
FOLD
5 Fold up the bottom edge a third time. Remember to just touch the base of each black tab above. Don't cover the tabs.
This is your third fold...
6 Tape shut between the tabs.
7 Open each tab to form a loop.
A PENCIL HELPS.
8 Write your name here.
CENTER
5 4 3 2 1 0 1 2 3 4 5

NAME: CLASS:

BUILD A MATH BALANCE (2)

1 Fold tape over the ends of a clothespin.

Pinch ends flat.

2 Cut out a narrow strip from the center of the tape.

3 Clamp the clothespin to the pull-tab on a pop can like this.

4 Lay your beam across a clothespin. Then push a straight pin through the exact center of the crossmarks.

5 Balance your beam on the top of the clothespin. To do this, rest the pin between the ears.

6 Bend out 2 paper clips just a little . . .

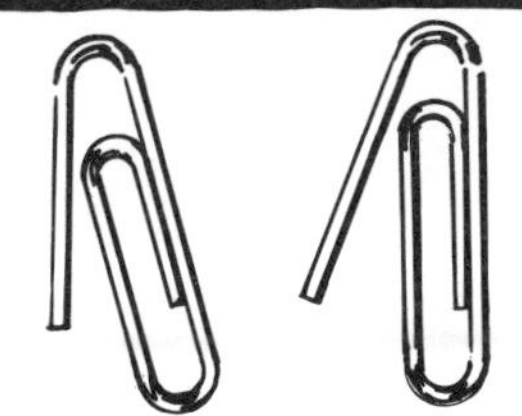

. . . then hang them from the outside loops like this.

7 Make the beam balance level by adding tape to the lighter (higher) side.

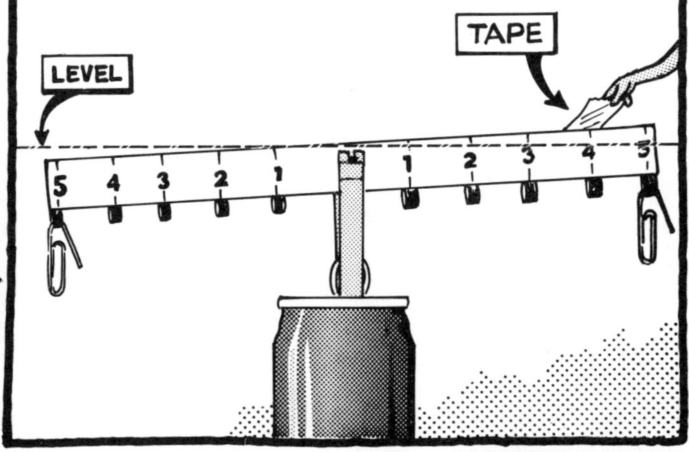

NAME: CLASS:

PAPER CLIP BALANCING

1 Pull out the arms on 15 paper clips just a little bit.

2 Start with a ***level*** beam. Then add paper clips to make it balance level again.

3 Draw 7 different ways to make your beam balance. Use an "X" for each paper clip.

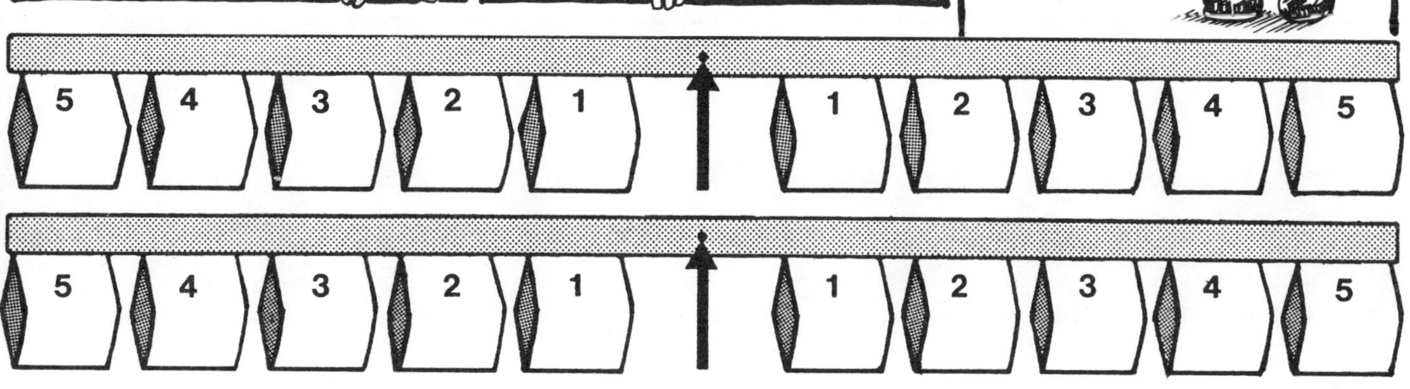

5 4 3 2 1 1 2 3 4 5

NAME: CLASS:

BALANCE ADDITION

1 Be sure your empty beam balances level.

2 Add ***just two*** paper clips to the right side to balance what is given on the left side.

3 Mark your two paper clip positions with an "X", then write an equation underneath. Make each way different.

ADD TWO CLIPS:

5 4 3 2 1 | 1 2 3 4 5

X (at 3)

3

ADD TWO CLIPS:

5 4 3 2 1 | 1 2 3 4 5

X (at 4)

4

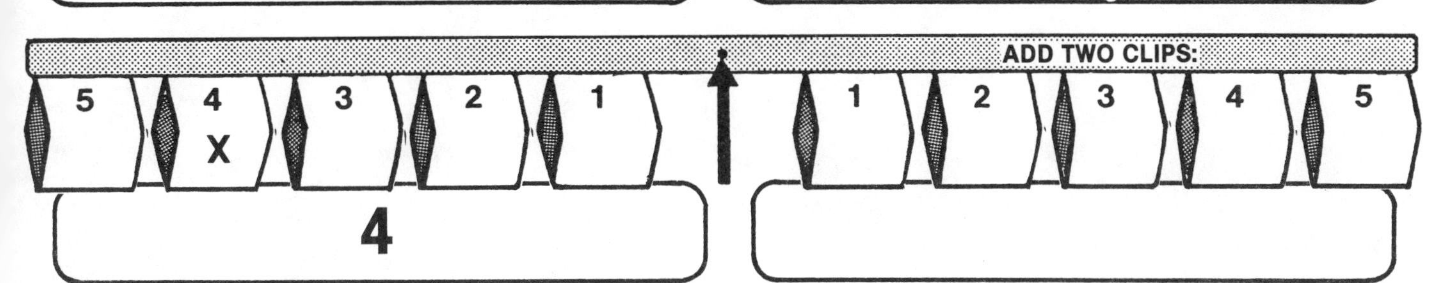

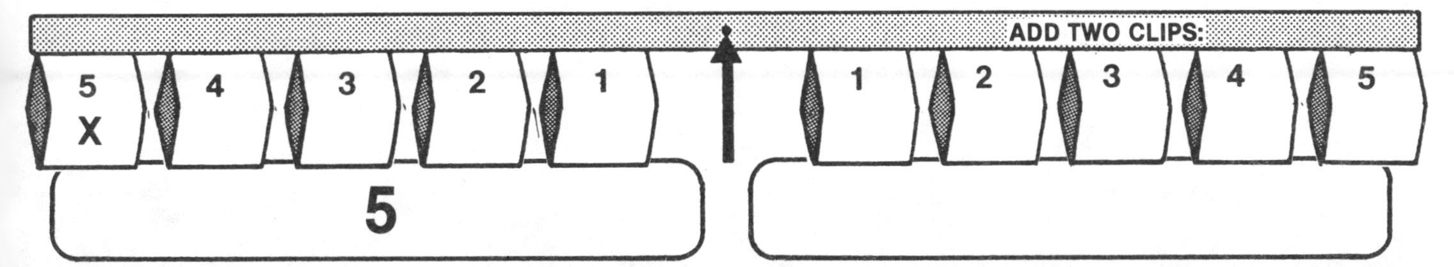

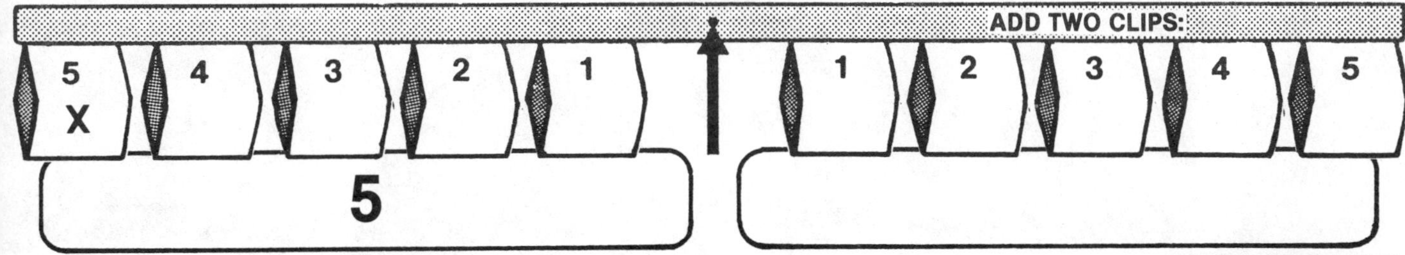

NAME: CLASS:

BALANCE MULTIPLICATION

1. Start with a balanced beam.
2. Add paper clips ***one at a time*** to either of the 2 given tabs until your beam balances.
3. Write a ***different*** multiplication equation in each box.

USE THE TABS SHOWN ON EACH BEAM BELOW.

5 | 3

4 | 5

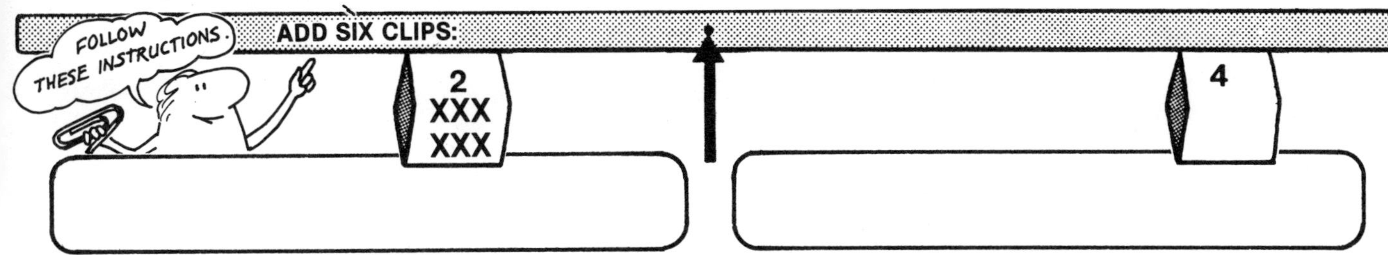

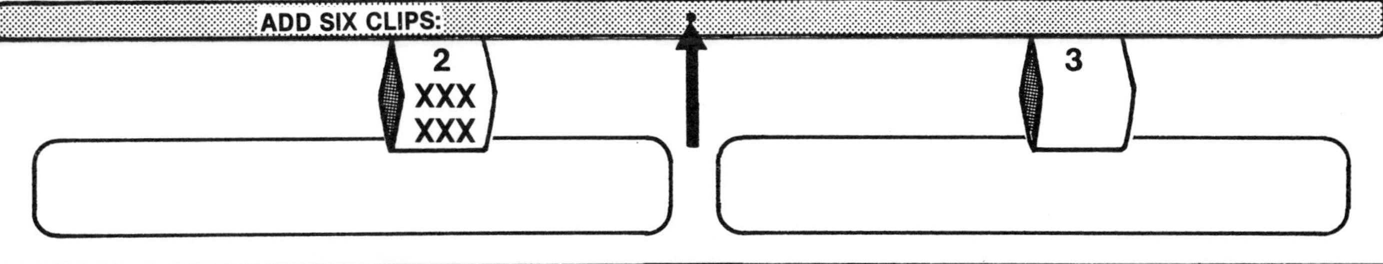

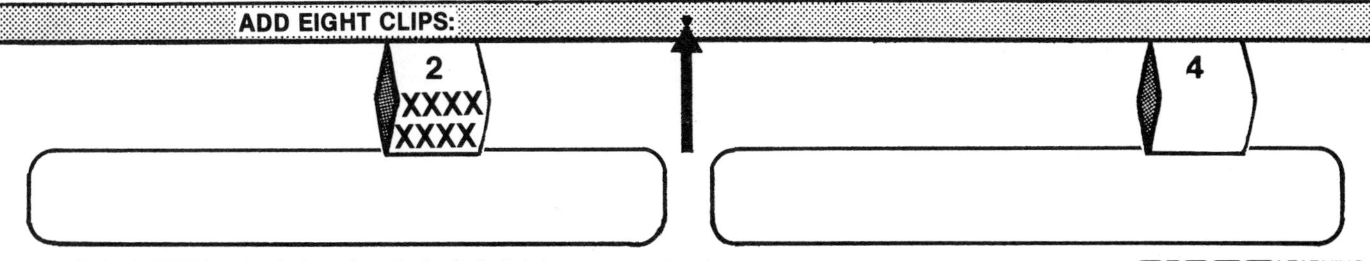

NAME: CLASS:

BALANCE PUZZLES

1 Add 2 paper clips to the left side of your beam as shown below:

2 Add ***just 3*** paper clips to the right side to make it balance.

MAKE EACH ANSWER DIFFERENT!

ADD THREE CLIPS:

5	4	3	2	1		1	2	3	4	5
XX										

ADD THREE CLIPS:

5	4	3	2	1		1	2	3	4	5
XX										

ADD THREE CLIPS:

5	4	3	2	1		1	2	3	4	5
XX										

ADD THREE CLIPS:

5	4	3	2	1		1	2	3	4	5
XX										

3 Add 2 paper clips to the left side of your beam as shown below.

4 Add ***just 4*** paper clips to the right side of the beam to make it balance.

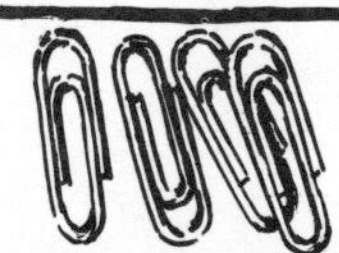

Make each way different.

ADD FOUR CLIPS:

5	4	3	2	1		1	2	3	4
XX									

ADD FOUR CLIPS:

5	4	3	2	1		1	2	3	4
XX									

ADD FOUR CLIPS:

5	4	3	2	1		1	2	3	4
XX									

ADD FOUR CLIPS:

5	4	3	2	1		1	2	3	4
XX									

NAME: CLASS:

MORE BALANCE PUZZLES

1 Add paper clips to the right side to balance what is given on the left side.

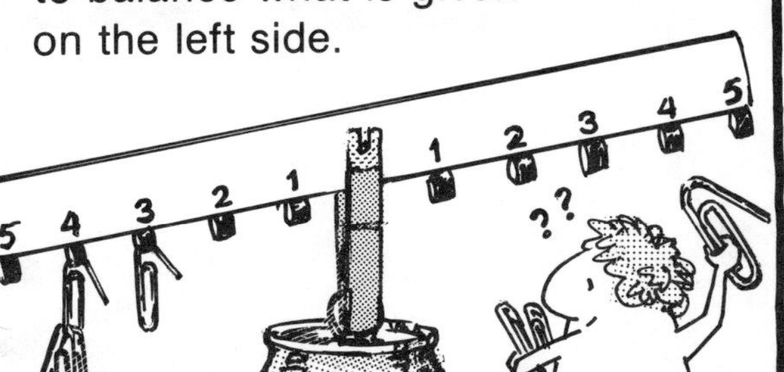

2 Write math equations in each pair of boxes to show that the beam balances.

Be sure to follow the instructions on the beams.

NO TAB MAY HAVE MORE THAN ONE CLIP.

5	4	3	2	1	↑	1	2	3	4	5
	XXX	X								

ADD CLIPS TO ONLY ONE TAB.

5	4	3	2	1	↑	1	2	3	4	5
	XX			X						

ADD TWO CLIPS TO TWO DIFFERENT TABS.

5	4	3	2	1	↑	1	2		4	5
	XX	X		XXX						

EACH TAB MUST HAVE THE SAME NUMBER OF CLIPS.

5	4	3	2	1	↑	1	2	3	4
XX XX									

NAME: CLASS:

Balancing( )8

DOES IT BALANCE?

1. Use math to decide if each beam balances as shown.

2. Write your ***prediction***. Will it balance?

3. Add the given paper clips to your beam. Do your results show your prediction is correct?

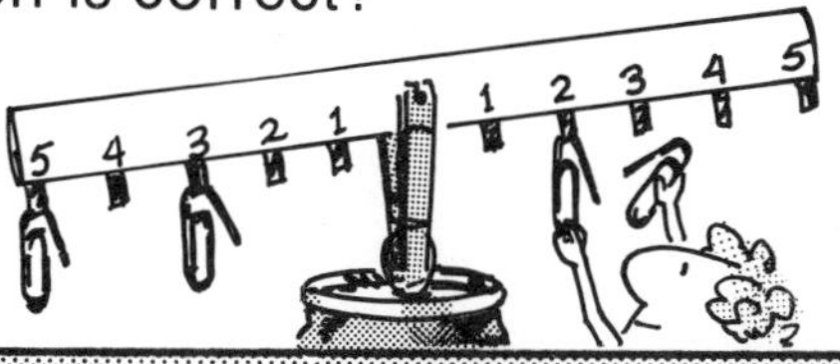

5	4	3	2	1	↑	1	2	3	4	5
X		X					XX	XX		

1. For your math:	2. Prediction:	3. Correct?	1. For your math:

5	4	3	2	1	↑	1	2	3	4	5
		XX XX				XX	X	X	X	

1. For your math:	2. Prediction:	3. Correct?	1. For your math:

5	4	3	2	1	↑	1	2	3	4	5
		XXX	XX	XX			X		XX	X

1. For your math:	2. Prediction:	3. Correct?	1. For your math:

5	4	3	2	1	↑	1	2	3	4	5
	XX	XX	XX				X		XX XX	

1. For your math:	2. Prediction:	3. Correct?	1. For your math:

NAME: CLASS:

Balancing( )9

WHICH WAY?

1 Use math to decide how each beam will tilt.

2 Write your ***prediction***. Will it tilt right, left, or will it balance?

3 Add the given paper clips to your beam.

Do your results show your prediction is correct?

5	4	3	2	1	↑	1	2	3	4	5
	X	X	XX				X		XX	

1. For your math: | **2. Prediction:** | **3. Correct?** | **1. For your math:**

5	4	3	2	1	↑	1	2	3	4	5
XX XX		XXX				X	X		XXX XXX	

1. For your math: | **2. Prediction:** | **3. Correct?** | **1. For your math:**

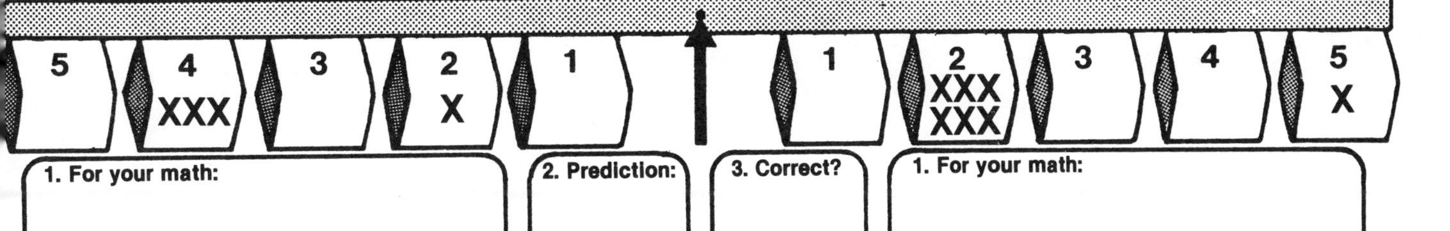

5	4	3	2	1	↑	1	2	3	4	5
	XXX		X				XXX XXX			X

1. For your math: | **2. Prediction:** | **3. Correct?** | **1. For your math:**

5	4	3	2	1	↑	1	2	3	4	5
X	X	XX	X	X			XX XXX		XX	

1. For your math: | **2. Prediction:** | **3. Correct?** | **1. For your math:**

NAME: CLASS:

SHORT'N'LONG ARM BALANCING

1 Pull the pin out of the center pivot.

2 Push it back through the mark above #1 to the *left* of zero.

3 Push this tab up *inside* the beam, out of the way.

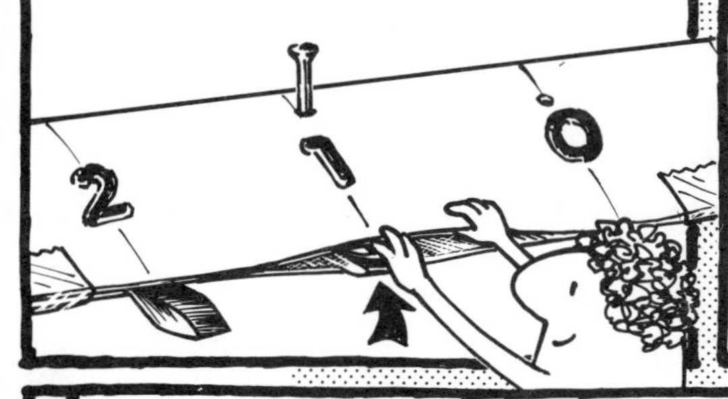

4 Put your off-center beam back on the clothespin.

Make it balance level by adding clay to the short arm.

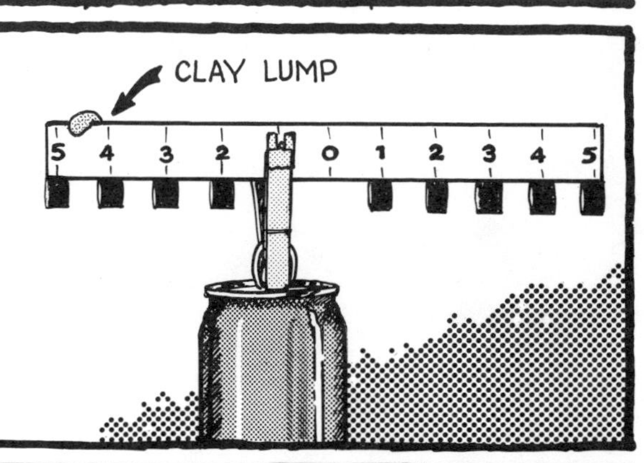

5 Here is one way to make 5 paper clips balance on your short n' long arm balance beam.

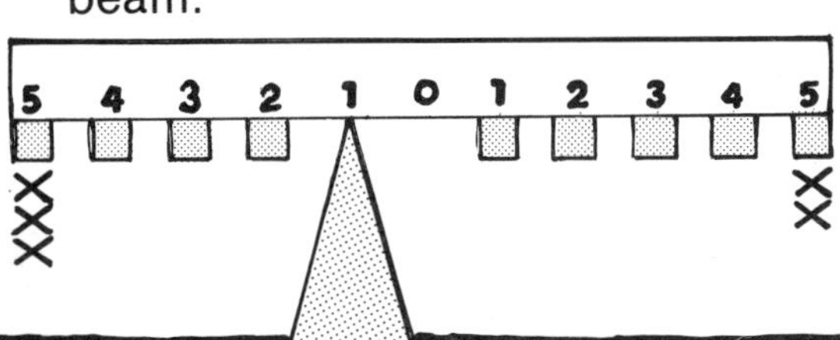

Show 3 other ways to make 5 paper clips balance.

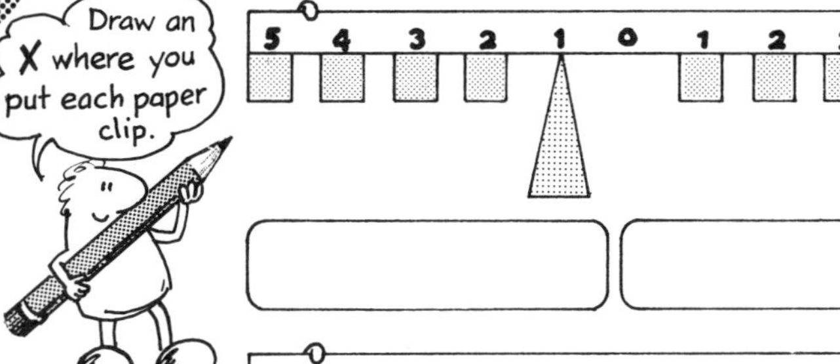

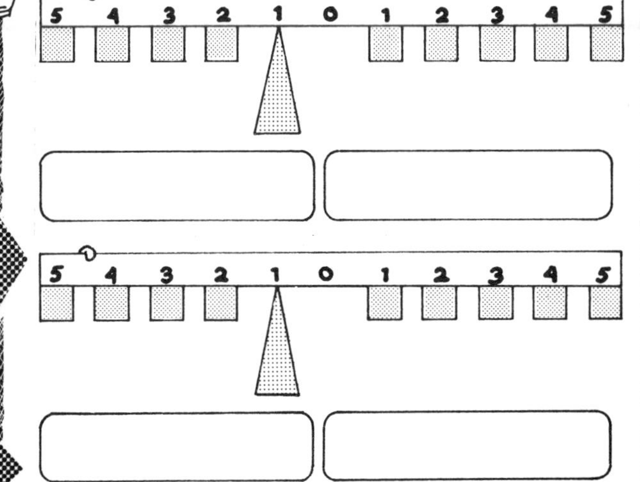

6 Your short n' long arm beam can still add and multiply if you renumber it. Begin at the new pivot and call it "0". Cross out the old numbers and write in the correct numbers on each tab.

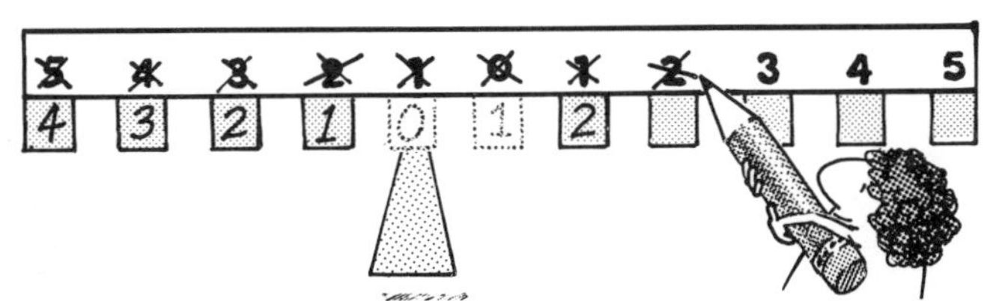

Now fill in each box, showing that these beams add and multiply correctly.

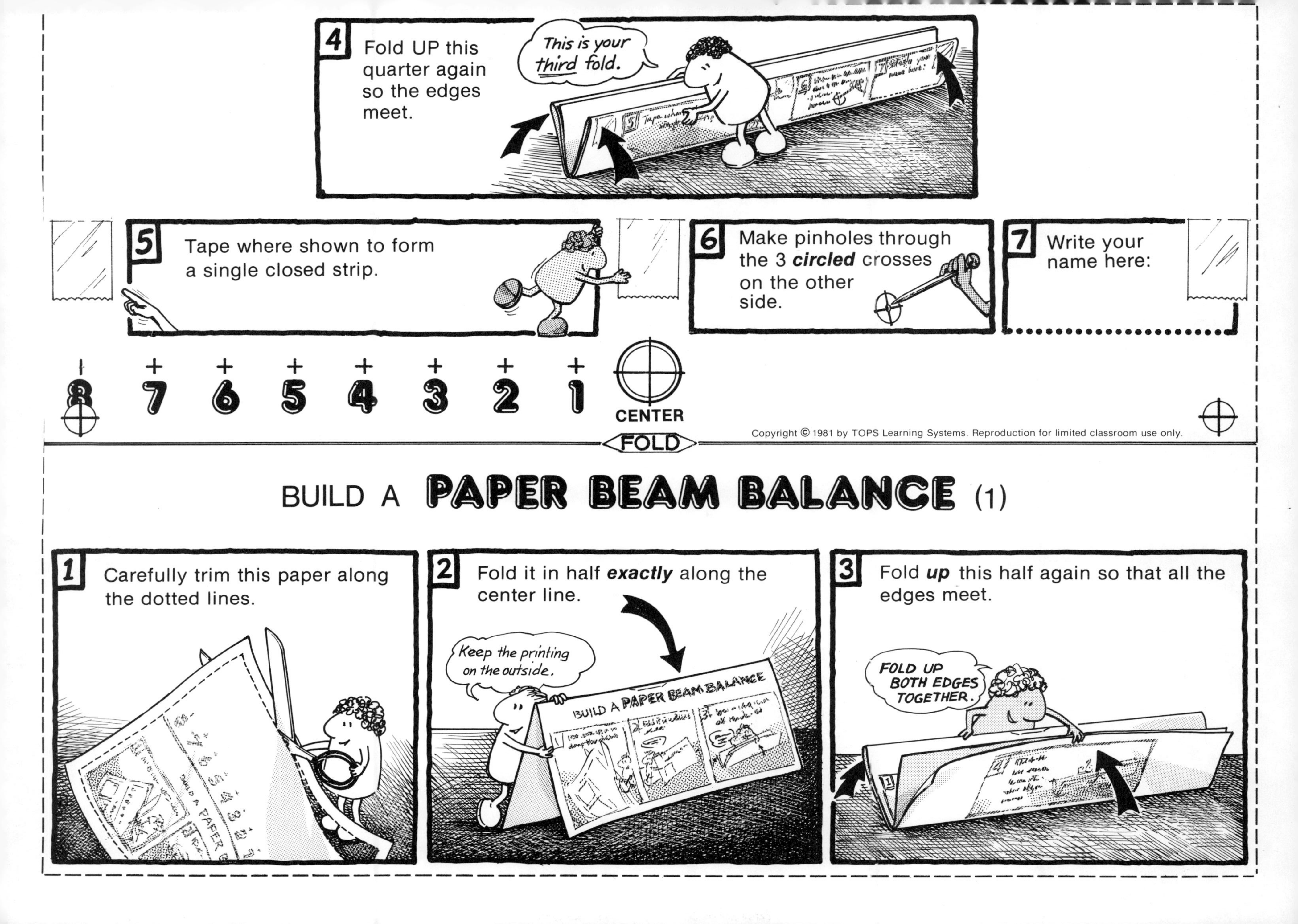

BUILD A PAPER BEAM BALANCE (1)
1 Carefully trim this paper along the dotted lines.
2 Fold it in half exactly along the center line.
Keep the printing on the outside.
3 Fold up this half again so that all the edges meet.
FOLD UP BOTH EDGES TOGETHER.
4 Fold UP this quarter again so the edges meet.
This is your third fold.
5 Tape where shown to form a single closed strip.
6 Make pinholes through the 3 circled crosses on the other side.
7 Write your name here:
8 7 6 5 4 3 2 1
CENTER
FOLD

NAME: CLASS:

BUILD A PAPER BEAM BALANCE (2)

1 Fold a 3 x 5 index card in half both ways. Cut the long fold to the center.

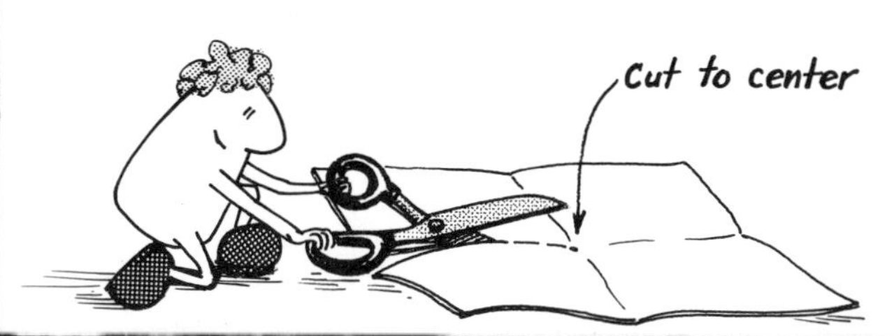

2 Fold up the tabs. Trim and tape.

3 Unbend a paper clip. Tape the small end to the outside of the folded-up card.

4 Cut into the opposite end about 1 cm (the width of a paper clip). Fold up and tape.

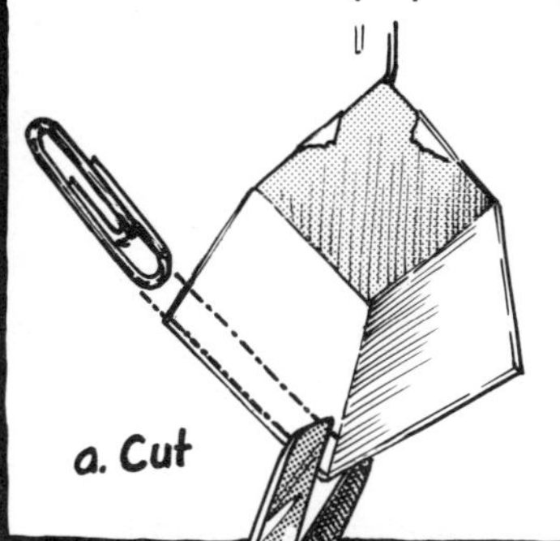

5 Repeat these steps to make a second weighing "pan".

6 Sharpen your pencil to a fine point. Use it to make the two ***end*** pinholes about as large as a pinhead.

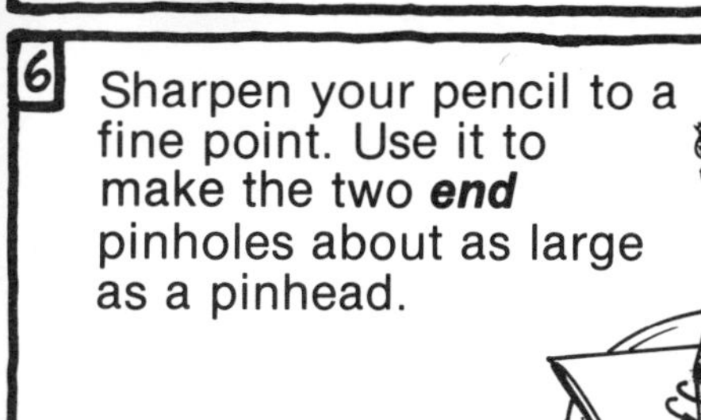

7 Poke 2 more unbent paper clips through the enlarged holes. Stick a pin through the center.

8 Bend the paper clip on each weighing pan forward, then loop the free end.

9 Balance your paper beam on a pin and taped clothespin as before. Hang a pan at each end, then center the beam with a clay rider.

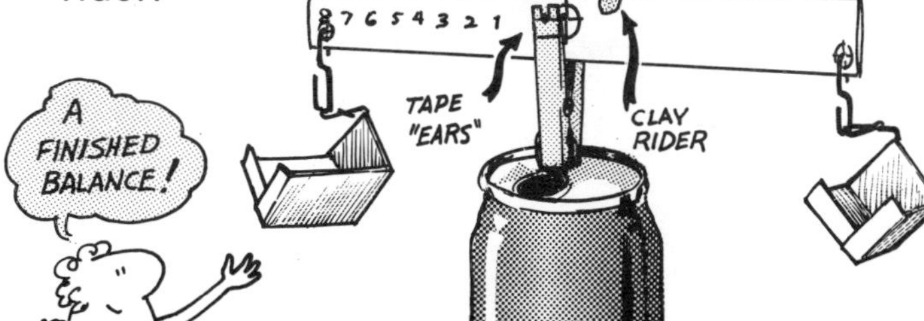

NAME: CLASS:

SEEDS & PAPER CLIPS

1 Center your beam. Add seeds and paper clips to fill in each box.

10 paper clips = ☐ pinto beans

10 pinto beans = ☐ popcorns

10 popcorns = ☐ lentils

10 lentils = ☐ rice

2 Fill in each box again. Don't weigh again. Just divide by 10!

1 paper clip = ☐ pinto beans

1 pinto bean = ☐ popcorns

1 popcorn = ☐ lentils

1 lentil = ☐ rice

3 Use math to solve the equations in each of these boxes . . .

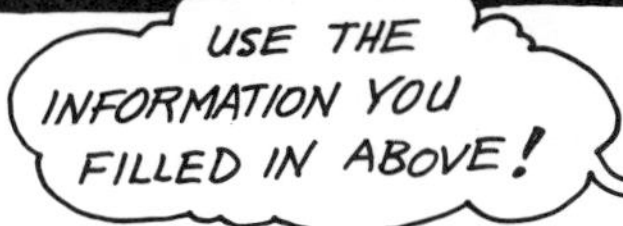

1 paper clip = ? popcorns	1 pinto bean = ? lentils	1 popcorn = ? rice
(a) Math Calculations:	(b) Math Calculations:	(c) Math Calculations:

. . . then check each answer on your balance.

(a) Balance Check:	(b) Balance Check:	(c) Balance Check:

NAME:

CLASS:

PAPER CLIP WEIGHING (1)

1 Weigh each item on your balance in paper clips.

You may write **PAPER CLIPS** like this...

1 PENNY = ... p.c.

1 SHEET OF PAPER = ...

1 WASHER = ...

1 INDEX CARD = ...

2 Weigh at least 4 other things in your classroom. Write your answer in paper clips.

1.

2.

3.

4.

Write on the back if you wish to weigh more items.

3 Find the weight of a rice grain in paper clips. Show your math.

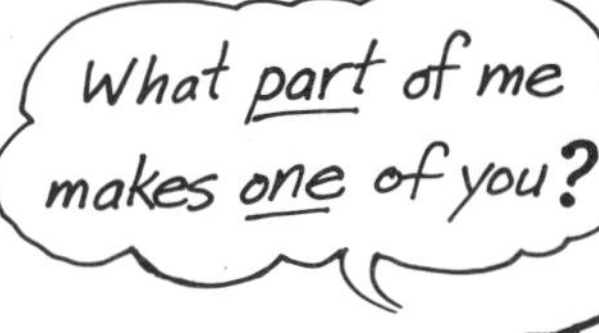

NAME: CLASS:

PAPER CLIP WEIGHING (2)

1 Carefully cut out the strip of lined paper at the bottom.

2 Cut parallel strips off the end until it weighs exactly 1 paper clip.

3 Divide this "paper clip strip" into 10 equal parts.

4 Cut your strip to make these paper clip fractions:

5 Weigh each item to the nearest tenth of a paper clip:

6 Compare these weights with what you found before.

NAME: CLASS:

Balancing( )16

PAPER SQUARE WEIGHTS

1 Cut out each group of squares along the ***dotted lines*** only.

DON'T cut the SOLID lines.

2 Weigh each item in these "paper squares". If you need to add less than 1 square, cut one in halves or quarters.

ITEM	WEIGHT
1 pinto bean	paper squares
½ index card	
10 popcorns	
10 rice grains	
1 paper clip	

3 Weigh a penny in "paper squares". Use ***only*** your own 28 squares and some clay. Tell how you did this.

4

PAPER SQUARES

CUT THE DOTTED LINES ONLY:

TOPS LEARNING SYSTEMS

NAME: CLASS:

HEAVY 'N' LIGHT WEIGHTS

1 Weigh these seeds in paper squares.

10 rice grains = ... paper squares

10 lentils = ...

10 popcorns = ...

2 Divide by 10 to find the weight of one seed.

1 rice grain = ... paper squares

1 lentil = ...

1 popcorn = ...

3 Predict which side will have more weight by filling in the table.

	LEFT PAN:	RIGHT PAN:	GREATER WEIGHT ?
a.	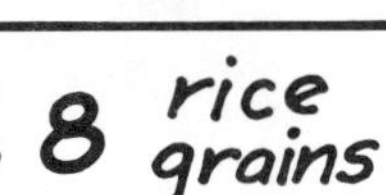 8 rice grains	 2 lentils	
b.	 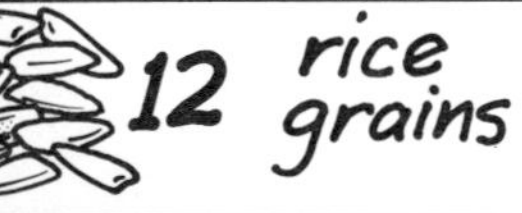 12 rice grains	 3 popcorns	
c.	  8 lentils	 2 popcorns	

a.

b.

c.

4 Check each prediction on your balance.

a. correct?

b. correct?

c. correct?

If you were wrong, try to find out why!

TOPS LEARNING SYSTEMS

NAME: CLASS:

SQUARES AND RECTANGLES

1 Carefully cut out the ruler ***plus*** each square and rectangle on the second sheet of paper.

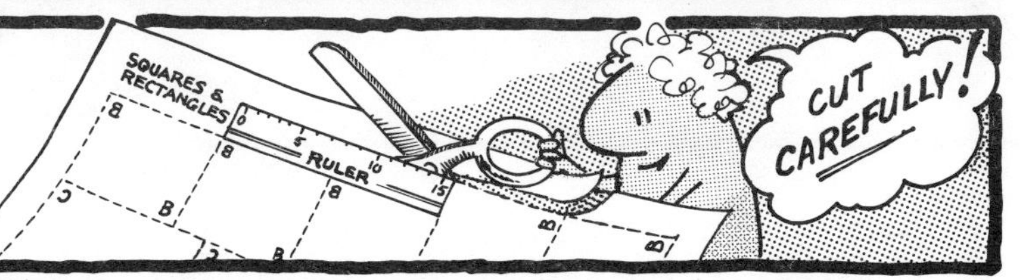

2 Move the clay rider so your balance is exactly level. Find out how many A's balance E's, etc.

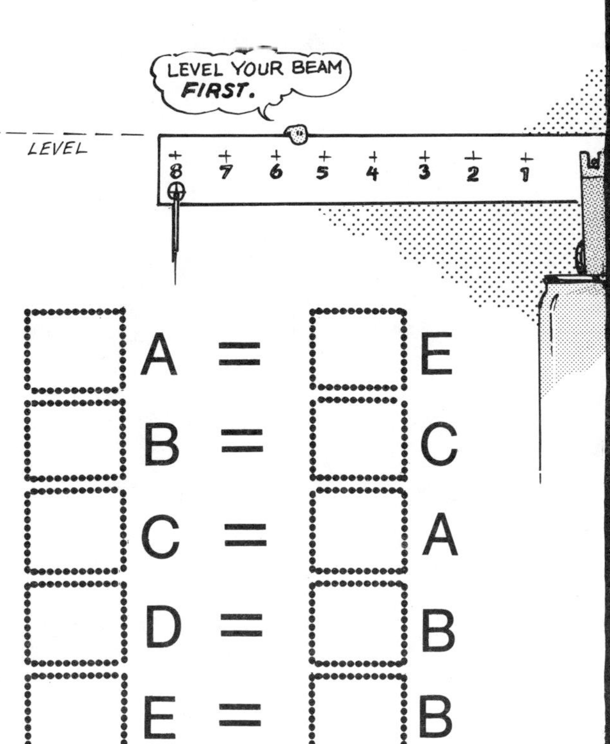

☐ A	=	☐ E	
☐ B	=	☐ C	
☐ C	=	☐ A	
☐ D	=	☐ B	
☐ E	=	☐ B	

3 Measure each piece with the ruler you cut out.

Then multiply to find area.

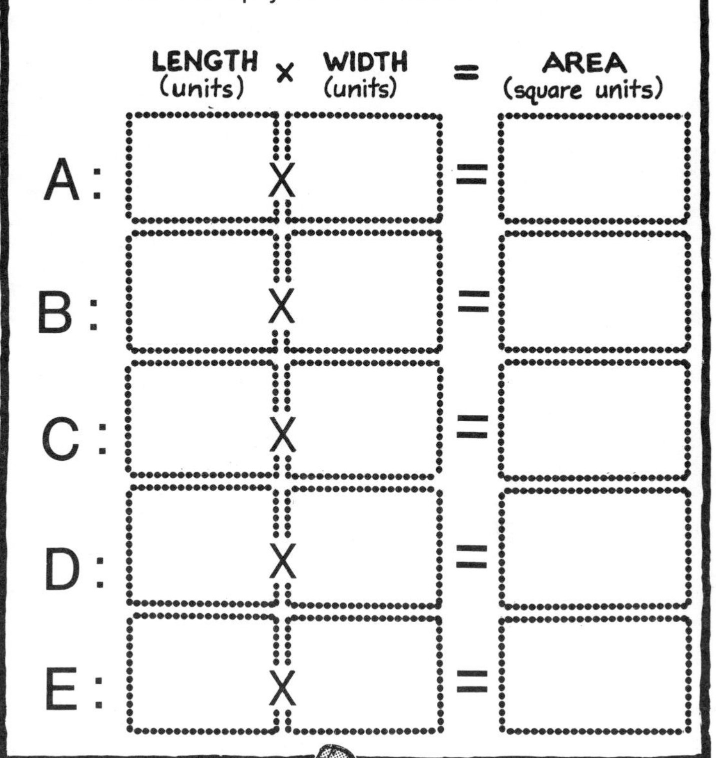

	LENGTH (units)	×	WIDTH (units)	=	AREA (square units)
A:		X		=	
B:		X		=	
C:		X		=	
D:		X		=	
E:		X		=	

4 Show that your ***areas*** in step 3 agree with your weight equations in step 2.

A	E	B	C	C	A	D	B	E	B

SQUARES AND RECTANGLES

0 5 Ruler 10 15

B B B B B
B B B B B
C C D A
C C D A
A
A
E E A A
A A
E E A A

NAME: CLASS:

EDUCATED GUESS

1 Start with 2 sheets of paper that are exactly the same size and weight.

2 Ask a friend to hide from 1 to 10 paper clips in one of the papers while you look away.

3 Use your balance to guess how many paper clips are wrapped inside.

My guess was ☐ correct ☐ wrong

4 Repeat this experiment until you guess the right amount at least 3 times in a row. Tell how you did it.

5 As a bank teller, your job is to count pennies to wrap in 50¢ rolls.

Tell how you would use a balance to make your job easier.

NAME: CLASS:

MOUNTAIN OF PAPER CLIPS

1 Punch pinholes through all 7 crossmarks along your beam. It's easy if you lay the beam on a styrofoam cup.

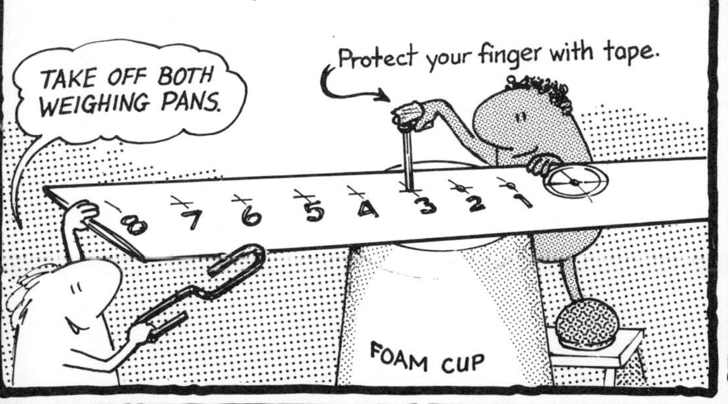

2 Hang a paper clip "hook" on the left side. Slide a regular clip over the right end so your beam balances level on its center pivot.

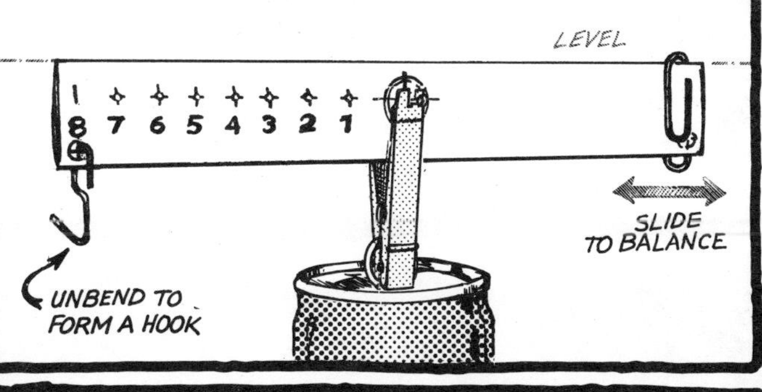

3 Fill in the data table: Find the number of paper clips you need to balance the beam at each new pivot position. Then graph your results in a smooth line.

DATA TABLE:

Number of clips to balance	Pivot Position
	0
	1
	2
	3
	4
	5
	6
	7

4 Could a "mountain" of paper clips hanging ***directly under*** pivot #8 raise the beam to a level position? Your graph is telling you the answer.

TOPS LEARNING SYSTEMS